A.F.M. WATSON.

A First Course on Complex Functions

CHAPMAN & HALL
MATHEMATICS SERIES

A First Course on Complex Functions

G. J. O. JAMESON

Lecturer in Mathematics
University of Warwick

CHAPMAN AND HALL LTD
11 NEW FETTER LANE LONDON EC4

First published 1970

Printed in Great Britain by
Spottiswoode, Ballantyne and Co. Ltd.
London and Colchester
SBN 412 09710 9

Distributed in the U.S.A.
by Barnes & Noble, Inc.

Contents

Preface

This book contains a rigorous coverage of those topics (and only those topics) that, in the author's judgement, are suitable for inclusion in a first course on Complex Functions. Roughly speaking, these can be summarized as being the things that can be done with Cauchy's integral formula and the residue theorem. On the theoretical side, this includes the basic core of the theory of differentiable complex functions, a theory which is unsurpassed in Mathematics for its cohesion, elegance and wealth of surprises. On the practical side, it includes the computational applications of the residue theorem. Some prominence is given to the latter, because for the more sceptical student they provide the justification for inventing the complex numbers.

Analytic continuation and Riemann surfaces form an essentially different chapter of Complex Analysis. A proper treatment is far too sophisticated for a first course, and they are therefore excluded.

The aim has been to produce the simplest possible rigorous treatment of the topics discussed. For the programme outlined above, it is quite sufficient to prove Cauchy's integral theorem for paths in star-shaped open sets, so this is done. No form of the Jordan curve theorem is used anywhere in the book.

The results of Complex Analysis are constantly motivated and illustrated by comparison with the real case. This policy is exemplified by the proof of Cauchy's integral theorem. Since similar formulae hold in the two cases for the integral of a derivative, it is enough to show that a differentiable function is itself a derivative. For a real function f, this would be done by showing that it is the derivative of F, where $F(x) = \int_a^x f$. The attempt to do the same in the complex case leads straight to the necessity of proving the Cauchy-Goursat theorem for a triangle.

The prerequisites are elementary Real Analysis and a little familiarity with metric spaces. The required metric space theory is set out on pages 1–7. This summary is strictly limited to results needed elsewhere in the book, and is not in any way intended as an alternative to the various existing texts on metric spaces. Though logically self-contained, this section is designed as a brisk revision rather than a beginner's introduction to metric spaces. Most, if not all, universities now include a course on metric or topological spaces quite early in their undergraduate programme, a policy which removes the necessity of beginning other courses with a lengthy discussion of preliminaries.

Chapter 2 contains the central part of the subject – the theory of functions that are differentiable on an open set. Digressions are kept to a minimum, in order not to detract from the beauty and unity of this classic sequence of results. The computational applications of the residue theorem are left to chapter 3, and all preparatory concepts, including integration, are fully discussed in chapter 1. Winding numbers are defined in order to state the residue theorem, but they are not permitted to interrupt the flow, on the grounds that all the usual applications of the residue theorem use paths that are easily seen to be simple. A proper discussion of winding numbers is deferred to chapter 3.

Chapter 1 may lack the unity and aesthetic appeal of chapter 2, but it would be too superficial to dismiss it as being merely introductory, or as being more typical of Real than Complex Analysis. Power series and the exponential function are really basic to Complex Analysis; it is debatable whether they are basic to Real Analysis. However, many of the results and proofs in chapter 1 have counterparts in the real case, and it should be possible to cover most of the material fairly quickly. Proofs are only omitted when the author is confident that all Real Analysis courses include them. For example, a proof is given of the theorem on differentiation of power series, but not of the rule for differentiating a product.

The section on the exponential and trigonometric functions should be viewed as a unit on its own. The properties of these functions are deduced from the power-series definitions, without assuming any results about the corresponding real functions. This procedure has the advantage of avoiding any logical gaps that might arise if a different definition had been used in a previous Real Analysis course. Omission of the proofs that do not involve complex numbers would result in an incomplete and disjointed account, without actually saving very much space.

We follow Ahlfors in defining paths to be functions defined on real intervals (and not equivalence classes of such functions). Integration is defined by approximating with step functions, since this serves to give some intuitive meaning to the integral, but it is made possible for the reader to opt for the purely formal definition of the integral as an integral of a function from $\mathbf{R}$ to $\mathbf{C}$.

Two slightly esoteric topics that are included both require an acknowledgement. I am indebted to Mr D. H. Fowler for showing me the method of evaluating the probability integral reproduced in 3.1, and to Prof. D. B. A. Epstein for drawing my attention to Palais' beautiful proof of the paucity of finite-dimensional division algebras (2.3, exercise 7). I am also indebted to the Editor of the Chapman and Hall Mathematics Series, Dr R. Brown, for reading the manuscript and making numerous useful suggestions.

Terminology and notation

The set of real numbers is denoted by **R**, and the set of complex numbers by **C**. Functions whose domain and range are both subsets of **R** are called *real functions*, and those whose domain and range are both subsets of **C** are called *complex functions*.

Terms and symbols relating to sets and mappings are used with their usual meaning. We write $f(x)$ for the image of x under the mapping f, and (occasionally) $f(A)$ for the set $\{f(x):x \in A\}$. Set-theoretic difference is denoted by \, and composition of functions by $\circ$. The closed real interval $\{x:a \leqslant x \leqslant b\}$ is denoted by $[a,b]$, and the open interval $\{x:a < x < b\}$ by (a,b). Other symbols introduced in the text are listed on pages 143–4.

Results (and only results) are numbered consecutively within sections, so that those in section 1.7 are 1.7.1, 1.7.2, etc. Only the most important ones are dignified with the appellation 'theorem'. Lemmas and corollaries are so designated, but all results not belonging to any of these categories are numbered without verbal qualification.

The derivative of the real function f at a (if it exists) is denoted by $f'(a)$, and the integral of f on $[a, b]$ by $\int_a^b f$. Similar notation is used for complex functions once the corresponding notions have been defined. It is, however, indisputably convenient to be able to use the classical notation d/dx and $\int \ldots dx$ on occasions. Its meaning can be made precise as follows. A function can be denoted either by giving it a name (such as f) or by the 'arrow' convention, e.g. $x \mapsto x^2$. The second method is often useful when dealing with particular functions, although it does not give any way of denoting the value of a function at a particular argument. When it is employed, we agree that the function f' can be denoted by $x \mapsto (d/dx)f(x)$. Here x, at its three occurrences, could be replaced by any other symbol not already used. With this understanding, it is correct to write (for example):

$$\frac{d}{dx}x^2 = 2x, \qquad \frac{d}{dt}t^2 = 2t.$$

In the case of integration, we agree that $\int_a^b f$ may be denoted by $\int_a^b f(x)dx$; again, another symbol could replace x.

Metric spaces

This section is a concise survey of the definitions and theorems applying to metric spaces that are required in this book. It is logically self-contained, but we make no pretence that it is a balanced account of the theory of metric spaces. For this, we refer to books on the subject, e.g. Simmons [6].

A **metric** on a set X is a mapping ρ from $X \times X$ to the non-negative real numbers satisfying the following conditions:

$$\rho(x,y) = 0 \qquad \text{if and only if } x = y,$$
$$\rho(y,x) = \rho(x,y),$$
$$\rho(x,z) \leqslant \rho(x,y) + \rho(y,z) \qquad \text{(the 'triangle inequality')},$$

for all x, y, z in X. An ordered pair (X,ρ), where ρ is a metric on X, is said to be a **metric space**. By restricting ρ, we obtain a metric on any subset of X.

Let X be a linear space (alias vector space) over the real field. A **norm** on X is a mapping $x \mapsto \|x\|$ from X to the non-negative real numbers satisfying the following conditions:

$$\|x\| = 0 \qquad \text{if and only if } x = 0,$$
$$\|\lambda x\| = |\lambda| . \|x\|,$$
$$\|x+y\| \leqslant \|x\| + \|y\|$$

for all x, y in X and λ in $\mathbf{R}$.

If $\| \ \|$ is a norm on a linear space X, then a corresponding metric ρ on X is defined by: $\rho(x,y) = \|x-y\|$. Every metric arising in this book is derived from a norm in this way.

0.1. *A norm is defined on* $\mathbf{R}^n$ *by:* $\|x\| = (x_1^2 + \cdots + x_n^2)^{1/2}$, *where* $x = (x_1, \ldots, x_n)$.

Proof. It is clear that $\|x\| = 0$ if and only if $x = 0$, and that $\|\lambda x\| = |\lambda| . \|x\|$ for λ in $\mathbf{R}$. To show that the triangle inequality holds, write

$$\langle x,y \rangle = x_1 y_1 + \cdots + x_n y_n.$$

Then $\langle x,x \rangle = \|x\|^2$ (note that this is non-negative for all x), and

$$\begin{aligned} \|x+y\|^2 &= \sum_{j=1}^{n} (x_j + y_j)^2 \\ &= \langle x,x \rangle + 2\langle x,y \rangle + \langle y,y \rangle, \end{aligned}$$

while

$$(\|x\| + \|y\|)^2 = \langle x,x \rangle + 2\|x\| . \|y\| + \langle y,y \rangle.$$

The result will follow if we can show that $|\langle x,y \rangle| \leqslant \|x\| . \|y\|$ for all x, y. Now for any real λ,

$$\begin{aligned} 0 &\leqslant \langle \lambda x - y, \lambda x - y \rangle \\ &= \lambda^2 \langle x,x \rangle - 2\lambda \langle x,y \rangle + \langle y,y \rangle. \end{aligned}$$

Since this is true for all real λ, we have $\langle x,y \rangle^2 \leqslant \langle x,x \rangle \langle y,y \rangle$, so that $|\langle x,y \rangle| \leqslant \|x\| . \|y\|$, as required.

In this book, $\mathbf{R}^n$ (and in particular, $\mathbf{R}$ and $\mathbf{R}^2$) will always be supposed to have the metric associated with the norm defined in 0.1.

We now list some definitions that are applicable to metric spaces.

Let (X, ρ) be a metric space. Let U be a subset of X, and let a be a point of X. Then U is a **neighbourhood** of a (and a is an **interior point** of U) if there exists $\delta > 0$ such that U contains the set $\{x \in X : \rho(x,a) < \delta\}$.

A sequence $\{x_n\}$ **converges** to a point a if, given a neighbourhood V of a, there is a corresponding integer N such that $x_n \in V$ whenever $n \geqslant N$. In symbols, we denote this statement by '$x_n \to a$ as $n \to \infty$' or by '$\lim_{n\to\infty} x_n = a$'.

A set G is **open** in (X,ρ) if it is a neighbourhood of each of its points. A set F is **closed** if $X \backslash F$ is open. A point a is in the **closure** of the set A if every neighbourhood of a meets A, or (equivalently) if a is the limit of a sequence of points of A. It is easily verified that a set is closed if and only if it is equal to its closure.

For any a in X and $r > 0$, it follows at once from the triangle inequality that $\{x : \rho(x,a) < r\}$ is open, and $\{x : \rho(x,a) \leqslant r\}$ is closed.

A point a is a **boundary point** of the set A if every neighbourhood of a meets both A and $X \backslash A$. Equivalently, a is in the closure of A and is not an interior point of A.

Let (X,ρ) and (Y,σ) be metric spaces, and let f be a mapping from X to Y. Then f is **continuous** at a point a of X if, given a neighbourhood V of $f(a)$, there is a neighbourhood U of a such that $f(U) \subseteq V$ (equivalently, $f^{-1}(V)$ is a neighbourhood of a). A function is **continuous on** X if it is continuous at each point of X.

0.2. *With the above notation, the following statements are equivalent:*

(i) *f is continuous on X;*

(ii) *for each open subset G of Y, $f^{-1}(G)$ is open in X;*

(iii) *for each closed subset F of Y, $f^{-1}(F)$ is closed in X.*

Proof. The equivalence of (ii) and (iii) is immediate on considering complements. Suppose that (i) holds, and let G be an open subset of Y. Take x in $f^{-1}(G)$, so that $f(x) \in G$. Then G is a neighbourhood of $f(x)$, so $f^{-1}(G)$ is a neighbourhood of x. Hence $f^{-1}(G)$ is open.

Now suppose that (ii) holds, and take x in X. Any neighbourhood V of $f(x)$ contains an open neighbourhood V'. Then $f^{-1}(V')$ is open, so is a neighbourhood of x. Hence f is continuous at x.

The mapping f is **uniformly continuous** on X if, given $\epsilon > 0$, there exists $\delta > 0$ such that $\sigma(f(x), f(y)) \leqslant \epsilon$ whenever $x, y \in X$ and $\rho(x,y) \leqslant \delta$.

The space (X,ρ) is **connected** if it is not expressible as the union of two disjoint, non-empty sets that are both open in X. When considering whether a subset A of X is connected, we must distinguish carefully between the concepts 'open in A' and 'open in X', the latter being a stronger condition.

Completeness

A sequence $\{x_n\}$ in a metric space (X,ρ) is said to be **Cauchy** if, given $\epsilon > 0$, there exists N such that $\rho(x_p, x_q) \leqslant \epsilon$ whenever $p, q \geqslant N$. Every convergent sequence is Cauchy. If, conversely, every Cauchy sequence is convergent, then (X,ρ) is said to be **complete**. It is a fundamental fact of Analysis that $\mathbf{R}$ is complete. We now show how to deduce from this that $\mathbf{R}^n$ is complete (we use a slightly different notation this time for elements of $\mathbf{R}^n$, to suit the needs of the case).

0.3. $\mathbf{R}^n$ *is complete. If* $\{x_k\}$ *is a sequence in* $\mathbf{R}^n$, *and* $x_k = (x_k(1), \ldots, x_k(n))$, *then* $x_k \to x$ *as* $k \to \infty$ *if and only if* $x_k(j) \to x(j)$ *as* $k \to \infty$ *for each* j.

Proof. We prove the second statement first. If $x_k \to x$, then $x_k(j) \to x(j)$, since

$$|x_k(j) - x(j)| \leqslant \|x_k - x\|.$$

Conversely, suppose that $x_k(j) \to x(j)$ for each j, and take $\epsilon > 0$. Then there is an integer k_0 such that whenever $k \geqslant k_0$, $|x_k(j) - x(j)| \leqslant \epsilon/n$ for $j = 1, \ldots, n$. Then $\|x_k - x\| \leqslant \epsilon$ for $k \geqslant k_0$, showing that $x_k \to x$ as $k \to \infty$.

Now suppose that $\{x_k\}$ is a Cauchy sequence in $\mathbf{R}^n$. Since $|x_p(j) - x_q(j)| \leqslant \|x_p - x_q\|$, we see that $\{x_k(j) : k = 1, 2, \ldots\}$ is a Cauchy sequence in $\mathbf{R}$ for each j. Since $\mathbf{R}$ is complete, this sequence has a limit in $\mathbf{R}$, say $x(j)$. By the above, $x_k \to x$.

Compactness

A metric space (X,ρ) is said to be **compact** if every family of open sets that covers X has a finite subfamily that covers X.

An immediate consequence that is often useful is the following: if for each point x of X, a neighbourhood $N(x)$ is given, then there is a finite subset $\{x_1, \ldots, x_n\}$ of X such that

$$X = \bigcup_{j=1}^{n} N(x_j).$$

By considering complements in the definition of compactness, we obtain at once the following criterion:

0.4. *The following statements are equivalent:*

(i) (X,ρ) *is compact;*

(ii) *If* $\mathscr{F}$ *is a family of closed subsets of* X *such that each finite subfamily of* $\mathscr{F}$ *has non-empty intersection, then* $\mathscr{F}$ *has non-empty intersection.*

0.5. *A compact subset of a metric space is closed.*

Proof. Suppose that A is not closed. Then there is a point x not in A such that every neighbourhood of x meets A. Let $G_n = \{y:\rho(x,y) > 1/n\}$. Then each G_n is open, and

$$A \subseteq \bigcup_{n=1}^{\infty} G_n.$$

If A could be covered by a finite number of the G_n, then, for some N, we would have $A \subseteq G_N$, so that $\{y:\rho(x,y) \leqslant 1/N\}$ would not meet A, contrary to hypothesis. Hence A is not compact.

0.6. *The continuous image of a compact metric space is compact. In other words, if* (X,ρ) *is a compact metric space, and* f *is a continuous mapping of* (X,ρ) *into a metric space* (Y,σ)*, then* $(f(X),\sigma)$ *is compact.*

Proof. Suppose that $f(X)$ is covered by a family $\mathscr{G}$ of open sets. Then X is covered by the sets $f^{-1}(G)$ $(G \in \mathscr{G})$, which are open, by 0.2. Therefore there is a finite subfamily $\mathscr{F}$ of $\mathscr{G}$ such that $f(X)$ is covered by $f^{-1}(G)$ $(G \in \mathscr{F})$. Then $f(X)$ is covered by $\mathscr{F}$.

0.7. *Let* (X,ρ) *be a compact metric space, and let* f *be a continuous mapping of* (X,ρ) *into a metric space* (Y,σ)*. Then* f *is uniformly continuous.*

Proof. Take $\epsilon > 0$. For each point a of X, there exists $\delta(a) > 0$ such that if $\rho(x,a) \leqslant \delta(a)$, then $\sigma(f(x),f(a)) \leqslant \epsilon/2$. There exist $a_1,\ldots, a_n$ such that X is contained in

$$\bigcup_{i=1}^{n} \{x:\rho(x,a_i) < \tfrac{1}{2}\delta(a_i)\}.$$

Let $\delta = \frac{1}{2} \min_{1 \leqslant i \leqslant n} \delta(a_i)$. Take x, x' in X such that $\rho(x,x') \leqslant \delta$. For some i, we have $\rho(x,a_i) \leqslant \frac{1}{2}\delta(a_i)$. For the same i, we have $\rho(x',a_i) \leqslant \delta(a_i)$. Hence $\sigma(f(x), f(x')) \leqslant \epsilon$.

A subset A of a normed linear space is said to be **bounded** if $\{\|a\| : a \in A\}$ is bounded in $\mathbf{R}$. The **diameter** of A is then defined to be

$$\sup\{\|a-b\| : a, b \in A\}.$$

It is elementary that if A is a bounded subset of $\mathbf{R}^n$, and $\epsilon > 0$ is given, then A can be covered by a finite family of sets, each of which has diameter less than ϵ (for instance, the covering sets can be taken to be suitably small cubes). The next result gives an important characterization of compact sets in $\mathbf{R}^n$.

0.8. *A subset of* $\mathbf{R}^n$ *is compact if and only if it is closed and bounded.*

Proof. (i) Suppose that A is compact. Then A is closed, by 0.5. The open sets $\{x : \|x\| < n\}$ $(n = 1,2,\ldots)$ cover A. Hence A is covered by a finite number of these sets (and, in fact, by one of them). In other words, A is bounded.

(ii) Suppose that A is closed and bounded, and that there is a family $\mathscr{G}$ of open sets that covers A and has no finite subfamily that covers A. Say that a set is *intractable* if it is not covered by any finite subfamily of $\mathscr{G}$. Express A as the union of a finite number of subsets, each of diameter less than 1. At least one of these sets, to be denoted by A_1, is intractable. Repeating the process, we construct an infinite sequence $\{A_n\}$ of intractable sets such that, for each n, $A_n \subseteq A_{n-1}$ and the diameter of A_n is less than $1/n$. For each n, choose a point x_n of A_n. If $m \geqslant n$, then x_m and x_n are both in A_n, so $\|x_m - x_n\| \leqslant 1/n$ (1). Hence $\{x_n\}$ is a Cauchy sequence, so has a limit x_0, by 0.3. Since A is closed, $x_0 \in A$. Also (letting $m \to \infty$ in (1)), $\|x_0 - x_n\| \leqslant 1/n$ for each n. Now x_0 is in some member G of $\mathscr{G}$. Since G is open, there exists N such that G contains $\{y : \|x_0 - y\| \leqslant 1/N\}$. For y in A_{2N}, we have

$$\|x_0 - y\| \leqslant \|x_0 - x_{2N}\| + \|x_{2N} - y\| \leqslant \frac{1}{2N} + \frac{1}{2N} = \frac{1}{N}.$$

Hence $A_{2N} \subseteq G$. But this contradicts the hypothesis that A_{2N} is intractable.

Combining this result (for $n = 1$) with 0.6, we have:

0.9 Corollary. *If (X, ρ) is a compact metric space, and f is a continuous mapping from X to* $\mathbf{R}$, *then $f(X)$ is closed and bounded in* $\mathbf{R}$. *In particular, $f(X)$ contains its supremum and infimum.*

An important special case of this corollary arises when f is the distance from a fixed point x_0, i.e. $f(x) = \rho(x, x_0)$ $(x \in X)$. It is elementary that f is continuous. 0.9 shows that a compact subset A of X contains a closest point to x_0.

CHAPTER ONE

Basic theory

This chapter is devoted to adapting the familiar concepts of Real Analysis to the complex case. The adaptation is often almost trivial, and at times we omit proofs that are formally identical to those applying in the real case. The main new feature is the possibility of defining paths in **C** (1.4) and of integrating along such paths (1.7). The reader with a sound knowledge of Real Analysis will not need to spend long on this chapter.

1.1. The complex number field

The real numbers **R** form a field in which every positive number has a unique positive square root (since the continuous function $x \mapsto x^2$ takes all its intermediate values). However, negative numbers have no square root in **R**. We shall construct a field which contains a subfield isomorphic to **R** and which has the property that every element has a square root. Later we shall be able to prove the much stronger result that every polynomial over this field has zeros.

Define two binary operations on $\mathbf{R}^2$ as follows:

$$(a_1, b_1) + (a_2, b_2) = (a_1 + a_2, b_1 + b_2),$$
$$(a_1, b_1)(a_2, b_2) = (a_1 a_2 - b_1 b_2, a_1 b_2 + b_1 a_2).$$

These operations will be called, respectively, **addition** and **multiplication**, and $\mathbf{R}^2$ with these operations will be called the **complex**

numbers and denoted by $\mathbf{C}$. Our definition of addition, of course, agrees with the usual 'linear-space' addition on $\mathbf{R}^2$. The reason for making this odd-looking definition of multiplication will become apparent below.

1.1.1 Theorem. $\mathbf{C}$ *is a field.*

Proof. Addition is clearly commutative and associative, has $(0,0)$ as identity and $(-a,-b)$ as inverse of (a,b).

Multiplication is commutative, since $(a_1,b_1)(a_2,b_2)$ and $(a_2,b_2)(a_1,b_1)$ are both equal to $(a_1a_2-b_1b_2, a_1b_2+b_1a_2)$. It is associative, since $(a_1a_2-b_1b_2, a_1b_2+b_1a_2)(a_3,b_3)$ and $(a_1,b_1)(a_2a_3-b_2b_3, a_2b_3+b_2a_3)$ are both equal to

$$(a_1a_2a_3-b_1b_2b_3-a_1b_2b_3-b_1a_2b_3,\\ a_1a_2b_3-b_1b_2b_3+a_1b_2a_3+b_1a_2a_3).$$

The element $(1,0)$ is an identity for multiplication. Now $(a,b)(a,-b) = (a^2+b^2,0)$. If $(a,b)\neq(0,0)$, then $a^2+b^2>0$, so (a,b) has multiplicative inverse

$$(a/(a^2+b^2), -b/(a^2+b^2)).$$

Since multiplication is commutative, there is only one distributive rule to verify. This is straightforward. Hence $\mathbf{C}$ is a field.

Define $h(a)=(a,0)$ for a in $\mathbf{R}$. Then h is a one-to-one mapping that 'preserves' addition and multiplication, i.e.

$$h(a+b)=h(a)+h(b), \qquad h(ab)=h(a)\,h(b)$$

for all a, b. It follows that $h(\mathbf{R})$ is a faithful copy of the field $\mathbf{R}$. We adopt the usual practice of identifying $\mathbf{R}$ with $h(\mathbf{R})$, writing a for $(a,0)$ (and, in particular, 0 for $(0,0)$). We shall allow ourselves to think of $\mathbf{R}$ as a subset of $\mathbf{C}$, and shall call members of $h(\mathbf{R})$ 'real'. Since $(c,0)(a,b)=(ca,cb)$, complex multiplication agrees with the linear-space definition of scalar multiplication when one of the factors is 'real'.

We also denote the complex number $(0,1)$ by i. Notice that

$$i^2=(0,1)(0,1)=(-1,0)=-1.$$

With the above convention, $bi = (0,b)$ for b in $\mathbf{R}$. Hence we can write the complex number (a,b) as $a+bi$, where a and b are 'real' (really the complex numbers $(a,0)$ and $(b,0)$). It is necessary to know the time-honoured terminology: a is called the 'real part' of $a+bi$, and b is called the 'imaginary part'. Needless to say, a is no more real, and no less imaginary, than b. Occasionally, however, it is useful to resort to the following notation:

$$a = \operatorname{Re}(a+bi), \qquad b = \operatorname{Im}(a+bi).$$

Given that we wanted to define a multiplication on a field containing $\mathbf{R}$ in such a way that -1 has a square root i, the field axioms give:

$$(a_1+b_1 i)(a_2+b_2 i) = (a_1 a_2 - b_1 b_2) + (a_1 b_2 + b_1 a_2) i.$$

This shows how we arrived at the definition of complex multiplication, and how one can deduce it every time one forgets it.

We show at once that $\mathbf{C}$ has the property we wanted of it, though this is a special case of later results:

1.1.2. *Every complex number has a complex square root.*

Proof. Take a non-zero complex number $z = a+bi$. We must find real numbers c, d such that $(c+di)^2 = a+bi$, that is:

$$c^2 - d^2 = a, \qquad 2cd = b.$$

Therefore we must have $b^2 = 4c^2(c^2-a)$, or $(2c^2-a)^2 = a^2+b^2$. Let $k = a + \sqrt{(a^2+b^2)}$. Then $k > 0$, and $(c+di)^2 = z$, where $c = \sqrt{(k/2)}$, $d = b/\sqrt{(2k)}$.

'Solving the quadratic' in the usual way, we deduce:

1.1.3 Corollary. *Every quadratic polynomial with complex coefficients has complex zeros.*

We mention that if $w, z \in \mathbf{C}$ and $w^2 = z$, then w and $-w$ are the only square roots of z. For if also $u^2 = z$, then

$$0 = w^2 - u^2 = (w+u)(w-u),$$

so $u = w$ or $u = -w$.

The question arises whether there is something special about $\mathbf{R}^2$, or whether it is possible to make $\mathbf{R}^n$ (for other n) into a field by defining multiplication suitably. It is reasonable to require also that the multiplication should be continuous. It is a remarkable fact, discovered by Frobenius in 1878, that this is not possible for any $n > 2$. Even if we drop the requirement that multiplication should be commutative, it is only possible for $n = 1$, 2 or 4. It is almost equally remarkable that the proof of this fact is facilitated by the study of complex functions; because of this, it appears in section 2.3 (exercise 7). The proof that it is possible in the case $n = 4$ is given in exercise 8 of the present section.

Moduli and conjugates

If $z = a + bi$, where a and b are real, we define the **modulus** $|z|$ of z to be the positive square root of $a^2 + b^2$ (in other words, the norm defined in 0.1), and the **conjugate** $\bar{z}$ of z to be $a - bi$. Of course, this definition of modulus agrees with that previously made for real numbers. The following relations, valid for all z, z_1, z_2 in $\mathbf{C}$, are immediate consequences of the definitions and of the fact that $|\;|$ is a norm:

$$\overline{z_1 + z_2} = \bar{z}_1 + \bar{z}_2, \qquad \bar{\bar{z}} = z, \qquad z\bar{z} = |z|^2,$$
$$1/z = \bar{z}/|z|^2 \; (z \neq 0), \qquad |\bar{z}| = |-z| = |z|,$$
$$|z_1 - z_2| \geqslant |z_1| - |z_2|.$$

Some further relations require slightly more proof:

1.1.4. *If z_1, z_2 are complex numbers, then:*
(i) $\overline{z_1 z_2} = \bar{z}_1 \bar{z}_2$
(ii) $|z_1 z_2| = |z_1| . |z_2|$.

Proof. (i) Let $z_k = a_k + b_k i$ $(k = 1, 2)$. Then

$$\begin{aligned}\bar{z}_1 \bar{z}_2 &= (a_1 - b_1 i)(a_2 - b_2 i)\\ &= (a_1 a_2 - b_1 b_2) - (a_1 b_2 + b_1 a_2) i\\ &= \overline{z_1 z_2}.\end{aligned}$$

(ii) $|z_1 z_2|^2 = (z_1 z_2)(\overline{z_1 z_2})$
$= z_1 z_2 \bar{z}_1 \bar{z}_2$, by (i),
$= (z_1 \bar{z}_1)(z_2 \bar{z}_2)$
$= |z_1|^2 |z_2|^2$.

1.1.5 Corollary. *If* z_1, z_2 *are complex numbers, and* $z_2 \neq 0$, *then*

$$\left|\frac{z_1}{z_2}\right| = \frac{|z_1|}{|z_2|}.$$

It must be remembered that no order relation has been defined on $\mathbf{C}$. A statement of the form $|z_1| < |z_2|$ (where $z_1, z_2 \in \mathbf{C}$) is, of course, an inequality of real numbers; no meaning has been attached to the formula $z_1 < z_2$. A phrase like 'if $r > 0$' will automatically be understood to mean 'if r is real and $r > 0$'.

We adopt the following as standard notation (where $a \in \mathbf{C}$ and $r > 0$):

$$D(a,r) = \{z \in \mathbf{C} : |z-a| < r\},$$
$$D'(a,r) = \{z \in \mathbf{C} : 0 < |z-a| < r\}.$$

Hence $D'(a,r) = D(a,r)\backslash\{a\}$. We refer to $D(a,r)$ as the *disc centre a, radius r*. If we need a name for $D'(a,r)$, we refer to it as the *punctured disc.*

It is important to be able to manipulate inequalities concerning the moduli of complex numbers. Sometimes a diagram may be helpful in interpreting a collection of inequalities, or in suggesting what one might expect (cf. exercises 1, 2). It should, however, be unnecessary to point out that diagrams never constitute a proof.

Example. If a, b, z are complex numbers such that $|b| < 1$, $z \neq \bar{a}$, and

$$\left|\frac{z-a}{z-\bar{a}}\right| \leqslant |b|,$$

then

$$|z| \leqslant |a| \frac{1+|b|}{1-|b|}.$$

For

$$\left|\frac{z-a}{z-\bar{a}}\right| = \frac{|z-a|}{|z-\bar{a}|},$$

so $|z-a| \leqslant |z-\bar{a}|.|b|$. But $|z-a| \geqslant |z|-|a|$, and $|z-\bar{a}| \leqslant |z|+|a|$, so $|z|-|a| \leqslant (|z|+|a|)\,|b|$, from which the statement follows.

Continuity and limits

We always regard **C** as being equipped with the metric ρ defined by $\rho(a,b) = |a-b|$. The various notions, such as continuity, that apply to metric spaces will therefore be defined, in the case of **C**, by reference to this metric.

If f and g are functions on a set A into **C**, we define the functions $f+g$ and fg 'pointwise', that is,

$$(f+g)(z) = f(z)+g(z), \qquad (fg)(z) = f(z)g(z) \qquad (z \in A).$$

The function $1/f$ is defined similarly if f does not take the value 0. In exactly the same way as for real functions, we can prove:

1.1.6. *Suppose that A is a subset of* **C**, *and that f, g are functions on A into* **C**. *If f and g are continuous at a point z_0 of A, then so are $f+g$ and fg. If, also, $0 \notin f(A)$, then $1/f$ is continuous at z_0.*

Constant functions and the identity function are clearly continuous, and from this and 1.1.6 it follows that all polynomials are continuous.

A notion closely allied to continuity is the following (also defined exactly as in the real case). Let f be a complex-valued function defined at all points of a neighbourhood of a point a of **C**, except possibly at a itself. Then we write '$f(z) \to l$ as $z \to a$' or '$\lim_{z\to a} f(z) = l$' for the following statement: given $\epsilon > 0$, there exists $\delta > 0$ such that $|f(z)-l| < \epsilon$ whenever $0 < |z-a| < \delta$. This is equivalent to saying that the function f_1 is continuous at a, where $f_1(z) = f(z)$ $(z \neq a)$, and $f_1(a) = l$.

The following notation is also useful. Let f be a complex-valued function defined on $\{z:|z|>r\}$ for some $r > 0$. Then we write

'$f(z) \to l$ as $z \to \infty$', or '$\lim_{z\to\infty} f(z) = l$' if, given $\epsilon > 0$, there exists $R > 0$ such that $|f(z) - l| < \epsilon$ whenever $|z| > R$.

Exercises 1.1

1 Prove that $|z+1| > |z-1|$ if and only if $\operatorname{Re} z > 0$.

2 If $\operatorname{Im} a > 0$ and $\operatorname{Im} b > 0$, show that

$$\left|\frac{a-b}{a-\bar{b}}\right| < 1.$$

Illustrate this statement by a diagram.

3 For any complex numbers a, b, prove that

$$|1-\bar{a}b|^2 - |a-b|^2 = (1-|a|^2)(1-|b|^2).$$

Deduce that if $|a| < 1$ and $|b| < 1$, then

$$\left|\frac{a-b}{1-\bar{a}b}\right| < 1.$$

4 Show that the mappings $z \mapsto |z|$, $z \mapsto \operatorname{Re} z$ and $z \mapsto \bar{z}$ are continuous on $\mathbf{C}$.

5 Let $f(x+iy) = xy/(x^2+y^2)$ for $x+iy \neq 0$. Does $f(z)$ tend to a limit as $z \to 0$?

6 If a, b, c are real and a, b are not both zero, prove that the distance from $(0,0)$ to the set $\{(x,y): ax+by+c=0\}$ is $|c|/\sqrt{(a^2+b^2)}$. Deduce that the distance from 0 to the line through z_1 and z_2 is

$$\frac{|\operatorname{Im}(z_1\bar{z}_2)|}{|z_1-z_2|}$$

7 (The algebraic terms appearing in this question are not used again in this book. The reader who is not acquainted with these terms may ignore this question with impunity). Suppose that X is a ring, and that addition and multiplication are defined on $X \times X$ in the same way as for the complex numbers.

Prove that $X \times X$ is a ring. By letting X be $\mathbf{C}$, show that $X \times X$ may have zero-divisors (and therefore fail to be a field) even when X is a field.

8 (The quaternions). Let $\mathbf{R}^4$ have its usual addition, and let $\{h, i, j, k\}$ denote its usual basis. Define a multiplication on $\mathbf{R}^4$ by stipulating the following conditions:

(i) h is the identity;
(ii) $i^2 = j^2 = k^2 = -h$;
(iii) $jk = -kj = i, \quad ki = -ik = j, \quad ij = -ji = k$;
(iv) multiplication is distributive with addition;
(v) $(\lambda x)y = x(\lambda y) = \lambda(xy)$ for λ in $\mathbf{R}$ and x, y in $\mathbf{R}^4$.

Prove that this addition and multiplication satisfy all the field axioms except commutativity of multiplication (so that $\mathbf{R}^4$ forms a *skew field* or *division ring*).

1.2. Sequences and series

Power series play a very important part in the theory of complex functions, because, as we shall see in chapter 2, every complex function that is differentiable on an open set is expressible locally as the sum of a power series. In this section we summarize the elementary theory of sequences and series of complex numbers and complex functions, leading up to the basic properties of power series. The results and proofs are formally identical to those applying in the real case.

By 0.3, $\mathbf{C}$ is complete, and convergence of a sequence $\{u_n + iv_n\}$ to $u + iv$ is equivalent to convergence of $\{u_n\}$ to u and $\{v_n\}$ to v. We omit the proof of the elementary statements listed in the next result.

1.2.1. *If $\{a_n\}$, $\{b_n\}$ are sequences in $\mathbf{C}$ that converge to a, b respectively, then:*

(i) $a_n + b_n \to a + b$,
(ii) $a_n b_n \to ab$,
(iii) *if* $a \neq 0$, *then* $1/a_n \to 1/a$.

Given a sequence $\{a_n\}$ in $\mathbf{C}$ such that $a_1 + \cdots + a_n \to s$ as $n \to \infty$, we say that the series $\sum_{n=1}^{\infty} a_n$ **converges to** s, and write

$\sum_{n=1}^{\infty} a_n = s$. Since $\mathbf{C}$ is complete, convergence of $\sum |a_n|$ implies convergence of $\sum a_n$, and we say that $\sum a_n$ is **absolutely convergent** in this case. The next result is the basic theorem on multiplication of series.

1.2.2 Theorem. *Suppose that $\{a_n\}$, $\{b_n\}$ $(n = 0, 1, 2, \ldots)$ are sequences of complex numbers such that*

$$\sum_{n=0}^{\infty} a_n = A, \qquad \sum_{n=0}^{\infty} b_n = B,$$

and $\sum_{n=0}^{\infty} |a_n|$ is convergent. Let

$$c_n = \sum_{k=0}^{n} a_k b_{n-k} \qquad (n = 0, 1, 2, \ldots).$$

Then $\sum_{n=0}^{\infty} c_n = AB$.

Proof. Let $\sum_{r=0}^{n} a_r = A_n$, $\sum_{r=0}^{n} b_r = B_n$, $\sum_{r=0}^{n} c_r = C_n$, and $B - B_n = d_n$. Then

$$\begin{aligned} C_n &= \sum \{a_r b_s : r + s \leqslant n\} \\ &= a_0 B_n + a_1 B_{n-1} + \cdots + a_n B_0 \\ &= A_n B - R_n, \end{aligned}$$

where

$$R_n = a_0 d_n + a_1 d_{n-1} + \cdots + a_n d_0.$$

Let $\alpha = \sum_{n=0}^{\infty} |a_n|$. Since $d_n \to 0$, $\{d_n\}$ is bounded: let $\beta = \sup\{|d_n| : n = 0, 1, 2, \ldots\}$. Given $\epsilon > 0$, there exists N such that $\sum_{n=N}^{\infty} |a_n| \leqslant \epsilon$ and $|d_n| \leqslant \epsilon$ for $n \geqslant N$. For $n \geqslant 2N$,

$$\begin{aligned} |R_n| &\leqslant (|a_0| + \cdots + |a_{n-N}|)\,\epsilon + (|a_{n-N+1}| + \cdots + |a_n|)\,\beta \\ &\leqslant (\alpha + \beta)\,\epsilon. \end{aligned}$$

Hence $R_n \to 0$. Also, $A_n B \to AB$, so $C_n \to AB$.

Sequences of functions and uniform convergence

If $\{f_n\}$ is a sequence of functions from A to $\mathbf{C}$, it is natural to consider the following condition (called 'pointwise convergence'): $f_n(a) \to f(a)$ for each a in A. Pointwise convergence, however, does not have very useful consequences. For instance, continuity of each f_n does not ensure continuity of f: if $A=[-1,1]$ and $f_n(x)=x^{2n}$, then $f(1)=f(-1)=1$ and $f(x)=0$ for $-1<x<1$.

Accordingly, we define a stronger condition on a sequence of functions. If f_n $(n=1,2,\ldots)$ and f are functions from A to $\mathbf{C}$, we say that $f_n \to f$ **uniformly** on A if, given $\epsilon>0$, there exists N such that whenever $n \geqslant N$, $|f_n(a)-f(a)| \leqslant \epsilon$ for all a in A.

Roughly speaking, this says that, for a given ϵ, the same N does for all a in A. The definition makes sense when A is any arbitrary set, but in this book it will always be a subset of $\mathbf{C}$. Uniform convergence is really just a case of convergence with respect to a metric, as we now show. Let $B(A,\mathbf{C})$ be the set of all bounded functions from A to $\mathbf{C}$. Then $B(A,\mathbf{C})$ is a linear space (with the natural 'pointwise' definitions of addition and scalar multiplication), and a norm $\| \ \|$ is defined on $B(A,\mathbf{C})$ by: $\|f\|=\sup\{|f(a)|:a\in A\}$. To say that $f_n \to f$ with respect to the metric associated with this norm means that $\|f_n - f\| \to 0$, and this is precisely what is meant by saying that $f_n \to f$ uniformly on A.

It would be instructive for the reader to convince himself directly that convergence is not uniform in the example above, though this follows from the next result.

1.2.3. *If each f_n is continuous on A, and $f_n \to f$ uniformly on A, then f is continuous on A.*

Proof. Take a in A and $\epsilon>0$. There exists n such that $|f_n(z)-f(z)| \leqslant \epsilon/3$ for all z in A. Since f_n is continuous at a, there exists $\delta>0$ such that $|f_n(z)-f_n(a)| \leqslant \epsilon/3$ whenever $z\in A$ and $|z-a|<\delta$. For such z, we have

$$|f(z)-f(a)| \leqslant |f(z)-f_n(z)|+|f_n(z)-f_n(a)|+|f_n(a)-f(a)|$$
$$\leqslant \epsilon.$$

We say that the series of functions $\sum f_n$ converges to the function s uniformly on A if the sequence of partial sums $(f_1+\cdots+f_n)$ $(n\geqslant 1)$ does so. The next result (known, quaintly, as the 'M-test'), gives a sufficient condition for this to happen.

1.2.4. *Let $\{M_n\}$ be a sequence of non-negative real numbers such that $\sum M_n$ is convergent, and let $\{f_n\}$ be a sequence of functions* *such that, for each n, $|f_n(z)| \leqslant M_n$ for all z in A. Then $\sum f_n$ is uniformly convergent on A.*

Proof. For each z in A, $\sum f_n(z)$ is convergent, since it is absolutely convergent. Let the sum be $s(z)$. Given $\epsilon > 0$, there exists N such that $\sum_{N+1}^{\infty} M_n \leqslant \epsilon$. For z in A and $p > N$,

$$\left|\sum_{N+1}^{p} f_n(z)\right| \leqslant \sum_{N+1}^{p} M_n \leqslant \epsilon,$$

Taking the limit as $p \to \infty$, we have

$$\left|\sum_{N+1}^{\infty} f_n(z)\right| = \left|s(z) - \sum_{n=1}^{N} f_n(z)\right| \leqslant \epsilon.$$

Power series

1.2.5 Theorem. *Let $\{a_n\}$ be a sequence of complex numbers. Then either* (i) *$\sum_{n=0}^{\infty} a_n z^n$ is absolutely convergent for all z in $\mathbf{C}$, or* (ii) *there exists $R \geqslant 0$ such that $\sum_{n=0}^{\infty} a_n z^n$ is absolutely convergent for $|z| < R$ and not convergent for $|z| > R$. If $0 \leqslant r < R$ (in case (i), if $r > 0$), then $\sum_{n=0}^{\infty} a_n z^n$ is uniformly convergent on $\{z : |z| \leqslant r\}$.*

Proof. Let $R = \sup\{|z| : \sum a_n z^n$ is convergent$\}$ (and put $R = \infty$ if this set is unbounded). Then $\sum a_n z^n$ is divergent for $|z| > R$, by definition. If $R > 0$, choose r such that $0 < r < R$. Then there exists z_0 such that $r < |z_0| < R$ and $\sum a_n z_0^n$ is convergent (if $R = \infty$, this holds for any $r > 0$). In particular, $\{|a_n z_0^n| : n = 0, 1, \ldots\}$ is bounded, say by M. Write $k = r/|z_0|$. Then, for $|z| \leqslant r$,

$$|a_n z^n| \leqslant |a_n| r^n = |a_n z_0^n| k^n \leqslant M k^n.$$

But $\sum M k^n$ is a convergent geometric series, so $\sum a_n z^n$ is uniformly convergent on $\{z : |z| \leqslant r\}$, by 1.2.4.

The number R in the theorem (and ∞ in case (i)) is said to be the **radius of convergence** of the power series $\sum a_n z^n$.

The next two corollaries follow without further proof from 1.2.2 and 1.2.3 respectively.

1.2.6 Corollary. *If $\sum_{n=0}^{\infty} a_n z^n$ and $\sum_{n=0}^{\infty} b_n z^n$ are both convergent for $|z| < R$, with sums $a(z)$, $b(z)$, and if*

$$c_n = \sum_{k=0}^{n} a_k b_{n-k} \qquad (n = 0, 1, 2, \ldots),$$

then $\sum_{n=0}^{\infty} c_n z^n = a(z)b(z)$ for $|z| < R$.

1.2.7 Corollary. *If $\sum_{n=0}^{\infty} a_n z^n$ converges to $s(z)$ for $|z| < R$, then s is continuous on $D(0, R)$.*

In particular, if $R > 0$ in 1.2.7, then $s(z) \to a_0$ as $z \to 0$.

Examples. (i) $\sum z^n$ is convergent for $|z| < 1$, not convergent for $|z| \geqslant 1$, since then $z^n \not\to 0$. Here $R = 1$, and the series diverges for all z such that $|z| = R$.

(ii) $\sum (1/n) z^n$ is convergent for $|z| < 1$, divergent for $z = 1$ (and hence for $|z| > 1$). Therefore $R = 1$ for this series too; note that the series converges when $z = -1$.

1.2.8 (The 'uniqueness theorem'). *Suppose that $\sum_{n=0}^{\infty} a_n z^n = a(z)$ and $\sum_{n=0}^{\infty} b_n z^n = b(z)$ for $|z| < R$, where $R > 0$. Suppose, also, that there are non-zero complex numbers $z_1, z_2, \ldots$ such that $z_k \to 0$ as $k \to \infty$ and $a(z_k) = b(z_k)$ for each k. Then $a_n = b_n$ for all n.*

Proof. Since a and b are continuous at 0, we have

$$\lim_{k\to\infty} a(z_k) = a_0, \qquad \lim_{k\to\infty} b(z_k) = b_0,$$

so $a_0 = b_0$. Suppose that it has been established that $a_r = b_r$ for $r = 0, 1, \ldots, m$. Let

$$a^*(z) = a_{m+1} + a_{m+2} z + \cdots, \qquad b^*(z) = b_{m+1} + b_{m+2} z + \cdots.$$

Then $a^*(z_k) = b^*(z_k)$ for each k, so, taking the limit as $k \to \infty$, we have $a_{m+1} = b_{m+1}$. This proves the result.

Exercises 1.2

1 Show that every convergent sequence in **C** is bounded.

2 If $a_n \to 0$, show that $(a_1 + \cdots + a_n)/n \to 0$.

3 Let $\{f_n\}$ be a sequence of functions $A \to \mathbf{C}$ that converges uniformly to f on A. If each f_n is uniformly continuous, prove that f is uniformly continuous.

4 Give an example of a power series that (a) converges for all z in $\mathbf{C}$, (b) converges only for $z = 0$.

5 Suppose that $\sum_{n=0}^{\infty} a_n z^n = s(z)$ for $|z| < R$, where $R > 0$.
(i) If $s(x)$ is real for all real x with $|x| < R$, show that each a_n is real.
(ii) If s is an even function (i.e. $s(-z) = s(z)$ for all z), show that $a_n = 0$ for all odd n.

6 Suppose that a_n is real and non-negative for each n, and that $\sum a_n R^n$ is convergent for some $R > 0$. Prove that $\sum a_n z^n$ is convergent for all complex z with $|z| = R$. Show by an example that this is not true if we drop the hypothesis that $a_n \geqslant 0$ for each n.

7 Show that $\sum z^n$ is not uniformly convergent on $D(0, 1)$.

8 For a sequence $\{u_n\}$ of real numbers, define

$$\limsup u_n = \inf_{r \geqslant 1} \sup_{n \geqslant r} u_n.$$

Show that the radius of convergence R of $\sum a_n z^n$ is given by $1/R = \limsup |a_n|^{1/n}$ (with obvious interpretations for $0, \infty$).

1.3. Line segments and convexity

This section is pure linear-space theory (though this fact could have been disguised by stating the results for $\mathbf{C}$ only). We include it here because it is often omitted from elementary courses on linear spaces.

Let X be a linear space over the real field. Then we make the following definitions:

(1) If $a, b \in X$, then the **line segment** $[a:b]$ is the set

$$\{\lambda a + (1 - \lambda) b : 0 \leqslant \lambda \leqslant 1\}.$$

(2) A subset A of X is **convex** if $[a:b] \subseteq A$ whenever $a, b \in A$.

(3) If $A \subseteq X$ and $a \in A$, then A is **star-shaped about** a if $[a:b] \subseteq A$ for all b in A.

We say simply that a set is **star-shaped** if it is star-shaped about some point. Clearly, a set is convex if and only if it is star-shaped about each of its points.

Examples. (i) Let a be a point of a normed linear space X, and let $r > 0$. Then $\{x \in X : \|x-a\| \leqslant r\}$ is convex. For if $\|x-a\| \leqslant r$, $\|y-a\| \leqslant r$ and $0 \leqslant \lambda \leqslant 1$, then

$$\begin{aligned} \|\lambda x + (1-\lambda)y - a\| &\leqslant \|\lambda(x-a)\| + \|(1-\lambda)(y-a)\| \\ &\leqslant \lambda r + (1-\lambda) r \\ &= r. \end{aligned}$$

(ii) Let T be the set of non-positive real numbers. We leave it to the reader to verify that $\mathbf{C}\backslash T$ is star-shaped about 1, but not convex.

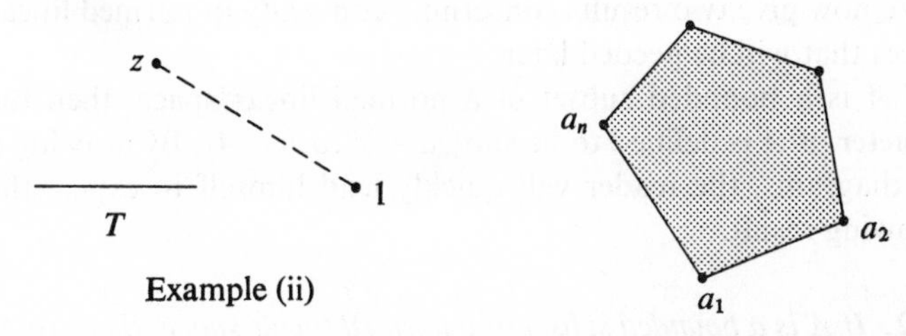

Example (ii)

Convex cover of a finite set

It is obvious that the intersection of any family of convex sets is convex, so that there is a smallest convex set containing a given set A, namely the intersection of all the convex sets containing A. This set is called the **convex cover** of A, and will be denoted by $\mathrm{co}(A)$. It can be described explicitly as follows. By a **convex combination** of points of A we mean a point of the form $\lambda_1 a_1 + \cdots + \lambda_n a_n$, where $a_i \in A$ and $\lambda_i \geqslant 0$ for each i, $\lambda_1 + \cdots + \lambda_n = 1$, and n is a positive integer.

1.3.1. co(A) *is the set of all convex combinations of points of* A.

Proof. Let E be the set of all convex combinations of points of A. Take x, y in E and α in $[0,1]$. Let

$$x = \lambda_1 a_1 + \cdots + \lambda_m a_m, \qquad y = \mu_1 b_1 + \cdots + \mu_n b_n.$$

Then $\alpha x + (1-\alpha)y \in E$, since

$$\alpha(\lambda_1 + \cdots + \lambda_m) + (1-\alpha)(\mu_1 + \cdots + \mu_n) = 1.$$

Hence E is convex. Now let F be any convex set containing A. We must show that F contains E. Suppose that F contains all convex combinations of not more than $n-1$ points of A, and let $x = \lambda_1 a_1 + \cdots + \lambda_n a_n$, where $n \geqslant 2$, $a_i \in A$ and $0 < \lambda_i < 1$ for each i. Let

$$\mu = \lambda_1 + \cdots + \lambda_n, \qquad y = \frac{1}{\mu}(\lambda_1 a_1 + \cdots + \lambda_{n-1} a_{n-1}).$$

Then $y \in F$ and $x = \mu y + \lambda_n a_n$. Hence $x \in F$, since F is convex. The result follows, by induction.

We now give two results concerning convexity in normed linear spaces that will be needed later.

If A is a bounded subset of a normed linear space, then the **diameter** of A is defined to be $\sup\{\|a-b\| : a, b \in A\}$. By drawing a few diagrams, the reader will quickly lead himself to expect the following result.

1.3.2. *If A is a bounded subset of a normed linear space, then* co(A) *is bounded, and has the same diameter as* A.

Proof. Let r be the diameter of A. It is sufficient to show that $\|x-y\| \leqslant r$ for all x, y in co(A). Take a_0 in A. The convex set $\{x : \|x-a_0\| \leqslant r\}$ contains A, so it contains co(A). That is, if $a \in A$ and $x \in$ co(A), then $\|x-a\| \leqslant r$. Now take x_0 in co(A). Then the convex set $\{y : \|y-x_0\| \leqslant r\}$ contains A, by the above, so it contains co(A). The required inequality follows.

1.3.3. *If A is a finite subset of a normed linear space, then* co(A) *is compact.*

Proof. Let $A = \{a_1, \ldots, a_n\}$. For $(\lambda_1, \ldots, \lambda_n)$ in R^n, define

$$f(\lambda_1, \ldots, \lambda_n) = \lambda_1 a_1 + \cdots + \lambda_n a_n.$$

By 1.3.1, f maps E onto $\text{co}(A)$, where $(\lambda_1, \ldots, \lambda_n) \in E$ if $\lambda_i \geqslant 0$ for each i and $\lambda_1 + \cdots + \lambda_n = 1$. It is elementary that E is closed and bounded. Hence, by 0.8, E is compact. We show that f is continuous; the result then follows, by 0.6. Take $\epsilon > 0$, and let $\alpha = \max\{\|a_i\| : 1 \leqslant i \leqslant n\}$. If $|\lambda_i - \mu_i| \leqslant \epsilon/\alpha n$ for each i, then

$$\left\| \sum_{i=1}^{n} (\lambda_i - \mu_i) a_i \right\| \leqslant \epsilon,$$

from which continuity of f follows.

Exercises 1.3

1 Prove that $\{x : \|x - a\| < r\}$ is convex.

2 Use 1.3.1 to give an alternative proof of 1.3.2.

3 Let A be a subset of a real linear space, and let a be a point of A. Describe the smallest set containing A that is star-shaped about a.

4 For subsets A, B of a real linear space, we write

$$A + B = \{a + b : a \in A \text{ and } b \in B\},$$
$$\lambda A = \{\lambda a : a \in A\} \qquad (\lambda \in \mathbf{R}).$$

Prove that (i) if A and B are convex, then so is $A + B$, and (ii) if A is convex and λ, $\mu > 0$, then $\lambda A + \mu A = (\lambda + \mu) A$.

1.4. Complex functions of a real variable

The first part of this section consists of straightforward adaptations of standard results on real functions, using the obvious correspondence between functions from $\mathbf{R}$ to $\mathbf{C}$ and pairs of functions from $\mathbf{R}$ to $\mathbf{R}$. The rest of the section is devoted to 'paths' – i.e. continuous mappings of real intervals into $\mathbf{C}$. These are needed when we come to integration of complex functions (section 1.7).

Differentiation and integration

If φ is a function from $\mathbf{R}$ (or part of $\mathbf{R}$) to $\mathbf{C}$, we write $\varphi = u + iv$ to mean that u and v are the real functions given by $\varphi(x) = u(x) + iv(x)$ $(x \in \mathbf{R})$. If $\lim_{h \to 0} (1/h)\,(\varphi(x+h) - \varphi(x))$ exists, we say that φ is **differentiable** at x, and denote the limit by $\varphi'(x)$. This limit clearly exists if and only if $u'(x)$ and $v'(x)$ both exist, and then $\varphi'(x) = u'(x) + iv'(x)$. If the domain of φ is an interval $[c,d]$, then the statement 'φ is differentiable on $[c,d]$' is to be interpreted as follows: $\varphi'(x)$ exists for $c < x < d$, and $\lim_{h \to 0^+} (1/h)(\varphi(c+h) - \varphi(c))$, $\lim_{h \to 0^-} (1/h)(\varphi(d+h) - \varphi(d))$ both exist. Denote these limits by $\varphi'(c)$, $\varphi'(d)$ respectively. If φ', defined in this way, is continuous on $[c,d]$, we say that φ is **smooth** on $[c,d]$.

The rule for differentiating a product can be deduced from the corresponding rule for real functions, or proved in the same way. The next result shows how the composition rule can be adapted.

1.4.1. *Let g be a function from $\mathbf{R}$ to $\mathbf{R}$, differentiable at a, and let φ be a function from $\mathbf{R}$ to $\mathbf{C}$, differentiable at $g(a)$. Then $\varphi \circ g$ is differentiable at a, with derivative $\varphi'(g(a))g'(a)$.*

Proof. Let $\varphi = u + iv$. Then $\varphi \circ g = u \circ g + i(v \circ g)$. By the composition rule for real functions, $(u \circ g)'(a) = u'(g(a))g'(a)$, and similarly for v. The result follows.

If $\varphi'(x) = 0$ for $c \leqslant x \leqslant d$, then $\varphi(d) = \varphi(c)$, by the mean value theorem applied to u and v. More generally, if φ is differentiable on $[c,d]$, then there exist ξ, η in (c,d) such that

$$\varphi(d) - \varphi(c) = (d-c)\,(u'(\xi) + iv'(\eta)).$$

Of course, we cannot necessarily assume that $\xi = \eta$.

There is also an obvious way to define integration: if $\varphi = u + iv$, we write

$$\int_c^d \varphi = \int_c^d u + i\int_c^d v$$

whenever the right-hand side exists. It is elementary that $\int_c^d \lambda\varphi = \lambda \int_c^d \varphi$ for any complex λ.

1.4.2. *Let* φ *be a continuous function from* $[c,d]$ *to* $\mathbf{C}$, *and let* $\Phi(x) = \int_c^x \varphi$ $(c \leqslant x \leqslant d)$. *Then* $\Phi'(x) = \varphi(x)$ *for* $c \leqslant x \leqslant d$.

Proof. Let $\varphi = u + iv$. Then $\Phi = U + iV$, where $U(x) = \int_c^x u$, $V(x) = \int_c^x v$. Since u and v are continuous, we have $U' = u$, $V' = v$.

1.4.3. *If* φ *is a smooth function from* $[c,d]$ *to* $\mathbf{C}$, *then* $\int_c^d \varphi' = \varphi(d) - \varphi(c)$.

Proof. This follows immediately from 1.4.2 and the mean value theorem.

1.4.4. *Suppose that* g *is a smooth function from* $[c,d]$ *to* $\mathbf{R}$, *and that* φ *is a continuous function from* $g[c,d]$ *to* $\mathbf{C}$. *Then*

$$\int_{g(c)}^{g(d)} \varphi = \int_c^d (\varphi \circ g) g'.$$

Proof. Let $\Phi(x) = \int_{g(c)}^x \varphi$ for x in $g[c,d]$. Then $\Phi'(x) = \varphi(x)$, by 1.4.2, and $(\Phi \circ g)' = (\varphi \circ g) g'$, by 1.4.1. Hence

$$\int_c^d (\varphi \circ g) g' = (\Phi \circ g)(d) = \int_{g(c)}^{g(d)} \varphi,$$

by 1.4.3.

The proof of the next result (which will be used repeatedly) requires slightly more ingenuity.

1.4.5. *If* φ *is a continuous function from* $[c,d]$ *to* $\mathbf{C}$, *then*

$$\left|\int_c^d \varphi\right| \leqslant \int_c^d |\varphi|.$$

Proof. $\int_c^d |\varphi|$ exists, since $|\varphi|$ is continuous. Let $\int_c^d \varphi = \lambda$, and suppose that $\lambda \neq 0$ (otherwise the result is clear). Let $\mu = \bar{\lambda}/|\lambda|$. Then $\int_c^d \mu\varphi = \mu\lambda = |\lambda|$, which is real, so $\int_c^d u = |\lambda|$, where u is the real part of $\mu\varphi$. Hence

$$|\lambda| \leqslant \int_c^d |u| \leqslant \int_c^d |\varphi|,$$

since $|u| \leqslant |\mu\varphi| = |\varphi|$.

Paths

A **path** in a metric (or topological) space X is a continuous function φ from a closed real interval $[c,d]$ into X. The range of φ is a compact subset of X, which we shall denote by φ^*. Intuitively, as t increases from c to d, $\varphi(t)$ describes the points of φ^* in some definite order. With this notation, $\varphi(c)$ is called the **initial point**, and $\varphi(d)$ the **final point**, of φ, and φ is said to be a path 'from $\varphi(c)$ to $\varphi(d)$'. If the initial and final points coincide, the path is said to be **closed**.

Equivalent paths. If φ is a path in X, with domain $[c,d]$, and g is a continuous, strictly increasing function from the real interval $[c',d']$ on to $[c,d]$, then $\varphi \circ g$ is a path in which the same points occur in the same order. Any path $\varphi \circ g$ arising in this way is said to be **equivalent** to φ. The reader will easily verify that this is an equivalence relation on the set of all paths in X. We shall see (usually trivially) that most of the definitions applying to paths are invariant under the relation of equivalence.

The **reverse** of a path $\varphi:[c,d] \to X$ is the path $-\varphi$ given by $(-\varphi)(t) = \varphi(-t)$ $(-d \leqslant t \leqslant -c)$. Roughly speaking, this is the same set of points described in the opposite direction.

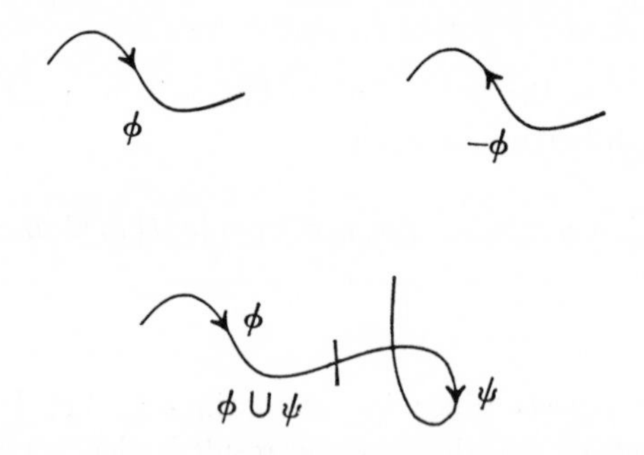

Joining of paths. Let φ and ψ be two paths in X, with respective domains $[c_1, c_2]$ and $[d_1, d_2]$, and such that $\varphi(c_2) = \psi(d_1)$. Let $\psi_1(t) = \psi(t + d_1 - c_2)$ for $c_2 \leqslant t \leqslant c_2 + d_2 - d_1$. Then ψ_1 is a path equivalent to ψ. Also, the function that agrees with φ on $[c_1, c_2]$ and with ψ_1 on $[c_2, c_2 + d_2 - d_1]$ is a continuous function from $[c_1, c_2 + d_2 - d_1]$ to X, in other words, a path in X. We denote this path by $\varphi \cup \psi$. Roughly speaking, it consists of the points of φ followed by those of ψ. Note that it is only defined when the final point of φ coincides with the initial point of ψ. Owing to the properties of equivalent paths, it is not usually necessary in practice to make the slightly clumsy change from ψ to ψ_1. In the same way, we can form the join of a finite sequence of paths, defining inductively

$$\varphi_1 \cup \cdots \cup \varphi_n = (\varphi_1 \cup \cdots \cup \varphi_{n-1}) \cup \varphi_n.$$

A path is said to be **piecewise-smooth** if it is of the form $\varphi_1 \cup \cdots \cup \varphi_n$, where each φ_j is smooth.

A simple example of a smooth path in **C** is the **directed line segment** from a to b. This is the path φ defined by:

$$\varphi(t) = a + t(b - a) = (1 - t)a + tb \qquad (0 \leqslant t \leqslant 1).$$

We denote it by $[a \to b]$. Note that $\varphi(0) = a$, $\varphi(1) = b$. Clearly, φ^* is the line segment $[a{:}b]$, as defined in 1.3. A path of the form $\varphi_1 \cup \cdots \cup \varphi_n$, where each φ_j is a directed line segment, will be called a **polygonal** path.

A metric (or topological) space is said to be **path-connected** if any two of its points can be joined by a path in the space.

Example. Every star-shaped subset of a normed linear space is path-connected. For if A is star-shaped about a, and $x, y \in A$, then the path $[x \to a] \cup [a \to y]$ lies in A.

From the fact that real intervals are connected, it is easy to prove that every path-connected space is connected (for the definition of *connected*, see page 3). The converse does not hold in general, but for open subsets of normed linear spaces, it certainly holds, as the next result shows.

1.4.6. *If G is connected, open subset of a normed linear space, then any two points of G can be joined by a polygonal path lying in G.*

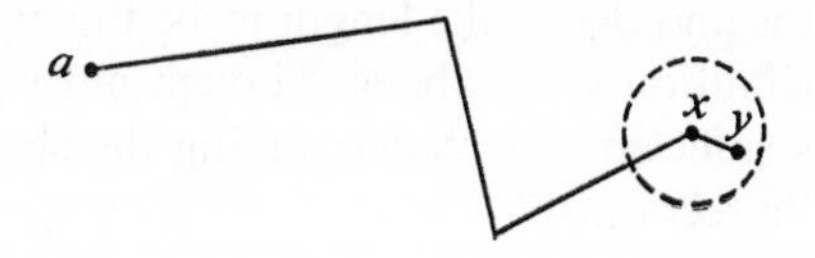

Proof. Choose a point a of G. Let A be the set of points of G that can be joined to a by a polygonal path lying in G. We show that A is both open and closed in G, from which it follows that $A = G$. Take x in A. There exists $\epsilon > 0$ such that the convex set $N_\epsilon(x) = \{y : \|y - x\| \leqslant \epsilon\}$ is contained in G. For y in $N_\epsilon(x)$, $[x:y] \subseteq G$, and hence $y \in A$. Therefore A is open.

Now let x be a point of G that is in the closure of A. Again, there exists $\epsilon > 0$ such that $N_\epsilon(x) \subseteq G$. But $N_\epsilon(x)$ contains a point of A, and hence $x \in A$.

Length of a path

Let φ be a path in **C**, with domain $[c,d]$. Take a finite subset S of (c,d). Let $t_1 < \cdots < t_{n-1}$ be the points of S in order, and write $c = t_0$, $d = t_n$. Define

$$L(\varphi, S) = \sum_{j=1}^{n} |\varphi(t_j) - \varphi(t_{j-1})|.$$

The triangle inequality shows that $L(\varphi, S') \geqslant L(\varphi, S)$ whenever $S' \supseteq S$. We say that φ is **rectifiable** (or of **bounded variation**) if

$$\{L(\varphi, S) : S \text{ a finite subset of } (c,d)\}$$

is bounded above, and define the supremum to be the **length** of φ, to be denoted by $L(\varphi)$.

Exactly the same finite sums are obtained for an equivalent path, so equivalent paths have the same length. It is also elementary that $L(-\varphi) = L(\varphi)$. We leave it as an exercise to prove that $L(\varphi \cup \psi) = L(\varphi) + L(\psi)$.

Example. If φ is the directed line segment $[a \to b]$, then $L(\varphi, S) = |b - a|$ for each S, so $L(\varphi) = |b - a|$.

The next theorem shows that the length of a smooth path can be expressed as an integral. It would be possible to restrict attention to smooth paths and define the length to be this integral, instead of using the definition given above. Though not very instructive, this method is sufficient for readers taking the shorter approach to integration in section 1.7.

1.4.7 Theorem. *If φ is a smooth path in* **C**, *with domain* $[c,d]$, *then* $L(\varphi) = \int_c^d |\varphi'|$.

Proof. For any finite subset S of (c,d), we have $\varphi(t_j) - \varphi(t_{j-1}) = \int_{t_{j-1}}^{t_j} \varphi'$, by 1.4.3, so $|\varphi(t_j) - \varphi(t_{j-1})| \leqslant \int_{t_{j-1}}^{t_j} |\varphi'|$, by 1.4.5. Hence $L(\varphi) \leqslant \int_c^d |\varphi'|$.

Take $\epsilon > 0$. Since φ' is uniformly continuous on $[c,d]$, there exists $\delta > 0$ such that if $x,\ y \in [c,d]$ and $|x-y| \leqslant \delta$, then $|\varphi'(x) - \varphi'(y)| \leqslant \epsilon$. Take $t_1 < t_2 < \cdots < t_{n-1}$ such that $t_j - t_{j-1} \leqslant \delta$ for each j (with $c = t_0$, $d = t_n$). Let $\varphi = u + iv$. For each j, there exist p_j, q_j in (t_{j-1}, t_j) such that

$$\varphi(t_j) - \varphi(t_{j-1}) = (t_j - t_{j-1})(u'(p_j) + iv'(q_j)).$$

Let $\psi(t) = u'(p_j) + iv'(q_j)$ for $t_{j-1} \leqslant t < t_j$. Then ψ is constant on each $[t_{j-1}, t_j)$, and $|\psi(t) - \varphi'(t)| \leqslant 2\epsilon$ for all t. Hence $\int_c^d (|\varphi'| - |\psi|) \leqslant 2\epsilon(d-c)$. Since $\int_c^d |\psi| = L(\varphi, S)$, the result follows.

Exercises 1.4

1 Let φ be a path in **C** with domain $[0,1]$. If $|\varphi(0)| < 1$ and $|\varphi(1)| > 1$, prove that φ^* contains a point with modulus 1.

2 Prove that the image of a path-connected space under a continuous mapping is path-connected.

3 Prove that $L(\varphi \cup \psi) = L(\varphi) + L(\psi)$.

1.5. Differentiation

Differentiation is defined in exactly the same way for complex functions as for real functions. Let f be a function from A to **C**, where $A \subseteq$ **C**, and let a be an interior point of A. We say that f is **differentiable** at a if

$$\lim_{h \to 0} \frac{1}{h}(f(a+h) - f(a))$$

exists (recall that the meaning of this was made precise at the end of 1.1). The limit is then called the **derivative** of f at a; we shall denote it by $f'(a)$.

Remembering that a function p is continuous at a if and only if $p(z) \to p(a)$ as $z \to a$, the proof of the following result becomes immediate:

1.5.1. *With the above notation, the following statements are equivalent:*

(i) *f is differentiable at a;*

(ii) *there exists a function $p: A \to \mathbf{C}$ that is continuous at a and satisfies*

$$f(z) = f(a) + (z - a)p(z) \qquad (z \in A).$$

We then have $f'(a) = p(a)$.

If G is an open subset of $\mathbf{C}$, then a function is said to be *differentiable on G* if it is differentiable at each point of G. Functions that are differentiable on the whole of $\mathbf{C}$ are called **entire** functions.

The derivative of a complex function cannot be interpreted in any convincing way as a 'gradient' – the definition is motivated by algebraic analogy with the real case rather than by geometric intuition.

Our present definition is also formally identical to the definition given in 1.4 of the derivative of a function from $\mathbf{R}$ to $\mathbf{C}$. It should be understood that, in our present definition, h is supposed to take all values in a *complex* neighbourhood of 0, whereas in 1.4 it was only supposed to take all values in a *real* neighbourhood.

Turning to properties of differentiable functions, we note first that a function is clearly continuous at any point where it is differentiable. We omit the proofs of the next three results, because they are exactly similar to the corresponding proofs for real functions (see e.g. Moss–Roberts [5], sect. 2.5). For comment on the d/dz notation, see page xii.

1.5.2. *If f and g are differentiable at a, then so are $f + g$ and fg, and*

$$(f+g)'(a) = f'(a) + g'(a),$$

$$(fg)'(a) = f'(a)g(a) + f(a)g'(a).$$

Hence $(d/dz)z^n = nz^{n-1}$ $(n = 1, 2, \ldots)$, and all polynomials are differentiable on $\mathbf{C}$.

1.5.3.

$$\frac{d}{dz}\frac{1}{z} = -\frac{1}{z^2} \qquad (z \neq 0).$$

In the next result, the domain A of f may be either a subset of $\mathbf{C}$ or a subset of $\mathbf{R}$; differentiability of f and $g \circ f$ is, of course, to be interpreted in the sense of this section or of 1.4 accordingly.

1.5.4. *Let f be a function from A to $\mathbf{C}$ that is differentiable at a, and let g be a function (with values in $\mathbf{C}$) that is defined on a neighbourhood of $f(a)$ and differentiable there. Then $g \circ f$ is differentiable at a, with derivative $g'(f(a))f'(a)$.*

We include the proof of the next result, though it is again identical to the proof applying to real functions:

1.5.5. *Suppose that A, B are open subsets of $\mathbf{C}$, and that f is a one-to-one function on A onto B, with inverse g. Take a in A, and let $f(a) = b$. Suppose that* (i) *$f'(a)$ exists and is non-zero, and* (ii) *g is continuous at b. Then $g'(b)$ exists, and equals $1/f'(a)$.*

Proof. By 1.5.1, there is a function $p: A \to \mathbf{C}$ that is continuous at a and satisfies $p(a) = f'(a)$ and

$$f(x) - f(a) = (x - a)p(x) \qquad (x \in A).$$

For y in B, we can substitute $g(y)$ for x in this equality to obtain

$$y - b = (g(y) - g(b))(p \circ g)(y) \qquad (y \in B).$$

Since $1/p \circ g$ is continuous at b, this shows that $g'(b)$ exists and equals $1/(p \circ g)(b) = 1/p(a) = 1/f'(a)$.

Differentiation of power series

The next theorem states that the derivative of the sum-function of a power series is the sum-function of the power series obtained by termwise differentiation. This result is essential for the later development of the theory, and it is important to understand why it needs proof (plausible though the statement may appear). The sum-function $\sum a_n z^n$ is a function associated, by a certain process, with a given sequence $\{a_n\}$, and it is by no means trivial that the function associated (by the same process) with the sequence $\{na_{n-1}\}$ is its derivative. In fact, the proof (which is again identical to the proof applying to the real case) uses most of the analysis developed so far. First we have a lemma:

1.5.6 Lemma. *If* $0 < t < 1$, *then* $nt^n \to 0$ *as* $n \to \infty$.

Proof. Let $1/t = 1 + u$. Then

$$\frac{1}{nt^n} = \frac{1}{n}(1+u)^n \geqslant \frac{n-1}{2}u^2 \qquad (n \geqslant 2),$$

by the binomial theorem. Hence $1/nt^n \to \infty$ as $n \to \infty$, from which the result follows.

1.5.7 Theorem. *Suppose that* $\sum_{n=0}^{\infty} a_n z^n = f(z)$ *for* $|z| < R$, *where* $R > 0$. *Then* $f'(z) = \sum_{n=1}^{\infty} na_n z^{n-1}$ *for* $|z| < R$.

Proof. Take z with $|z| < R$, and r, s such that $|z| < r < s < R$. For complex h such that $0 < |h| < r - |z|$, we have

$$\frac{1}{h}((z+h)^n - z^n) = (z+h)^{n-1} + (z+h)^{n-2}z + \cdots + z^{n-1}.$$

Hence

$$\frac{1}{h}(f(z+h) - f(z)) = \sum_{n=1}^{\infty} g_n(h),$$

where

$$g_n(h) = a_n((z+h)^{n-1} + (z+h)^{n-2}z + \cdots + z^{n-1}) \qquad (h \neq 0).$$

Defining $g_n(0)$ to be $na_n z^{n-1}$, we have that g_n is continuous for all h, and that $|g_n(h)| \leqslant n|a_n| r^{n-1}$ for $|h| \leqslant r - |z|$.

By 1.5.6, $n(r/s)^n \to 0$ as $n \to \infty$, so for all sufficiently large n, we have $n(r/s)^n \leqslant r$, or $nr^{n-1} \leqslant s^n$. Since $\sum_{n=1}^{\infty} |a_n| s^n$ is convergent, it follows that $\sum_{n=1}^{\infty} n|a_n| r^{n-1}$ is convergent. Therefore by 1.2.4 (the 'M-test'), $\sum_{n=1}^{\infty} g_n$ is uniformly convergent on $\{h: |h| \leqslant r - |z|\}$, and by 1.2.3, its sum is continuous at 0. Hence

$$\frac{1}{h}(f(z+h) - f(z)) \to \sum_{n=1}^{\infty} na_n z^{n-1} \qquad \text{as } h \to 0.$$

Notice that this theorem strengthens 1.2.7, since a differentiable function is continuous.

Putting $z = 0$ in 1.5.7, we obtain $f'(0) = a_1$, and, more generally, $f^{(n)}(0) = n!a_n$. This identity shows how the coefficients in a power series can be derived from the sum-function; the fact that they are determined by the sum-function follows, of course, from the 'uniqueness theorem' 1.2.8.

Theorem on variation

We now come to matters in which real and complex functions diverge. If f is a real function that is differentiable on a closed interval $[a,b]$, then the mean-value theorem states that $f(b)-f(a) = (b-a)f'(\xi)$ for some ξ between a and b. An important corollary states that if $f'(x) = 0$ for $a \leqslant x \leqslant b$, then f is constant on $[a,b]$. The difficulty in adapting the mean-value theorem to complex functions lies in interpreting the word 'between'. If we take it to mean 'on the line segment between', the statement becomes false (see exercise 6). However, the following weaker variant of the mean-value theorem is still sufficient to yield the corollary mentioned above: if $|f'(x)| \leqslant M$ for $a \leqslant x \leqslant b$, then $|f(b)-f(a)| \leqslant M(b-a)$. We now show that this statement generalizes successfully to complex functions. The real interval $[a,b]$ can be replaced, not only by a line segment, but by a smooth (or piecewise-smooth) path. Let us say that a complex function f is **smooth** on an open set if f' is continuous there (actually, we shall see in chapter 2 that the mere existence of f' on an open set implies its continuity).

1.5.8 Theorem. *Let φ be a piecewise-smooth path in* $\mathbf{C}$ *from a to b. Suppose that f is a smooth complex function defined on an open set containing φ^*, and that $|f'(z)| \leqslant M$ for all z in φ^*. Then*

$$|f(b)-f(a)| \leqslant ML(\varphi).$$

Proof. It is clearly sufficient to prove the theorem for smooth φ. Let the domain of φ be $[c,d]$, so that $\varphi(c) = a$, $\varphi(d) = b$. By 1.5.4, we have $(f \circ \varphi)' = (f' \circ \varphi)\varphi'$. This function is continuous on $[c,d]$, so 1.4.3 applies to show that

$$f(b)-f(a) = \int_c^d (f \circ \varphi)'.$$

Also, $|(f \circ \varphi)'(t)| \leqslant M|\varphi'(t)|$ for $c \leqslant t \leqslant d$. Hence

$$|f(b)-f(a)| \leqslant M\int_c^d |\varphi'| \qquad \text{(by 1.4.5)}$$
$$= ML(\varphi) \qquad \text{(by 1.4.7).}$$

1.5.9 Corollary. *Suppose that f is a smooth complex function defined on an open set containing $[a:b]$, and that $|f'(z)| \leqslant M$ for all z in $[a:b]$. Then $|f(b)-f(a)| \leqslant M|b-a|$.*

1.5.10 Corollary. *If G is a connected, open subset of $\mathbf{C}$, and f is a complex function such that $f'(z) = 0$ for all z in G, then f is constant on G.*

Proof. Take a, b in G. By 1.4.6, there is a polygonal path lying in G and joining a to b. By 1.5.8, $f(b) = f(a)$.

Differentiability on $\mathbf{C}$ and on $\mathbf{R}^2$

For functions defined on $\mathbf{R}^2$, there is a standard definition of 'differentiability' that makes no use of the multiplication that turns $\mathbf{R}^2$ into the complex numbers. Functions satisfying this definition will be said to be differentiable 'as functions on $\mathbf{R}^2$', while functions satisfying the definition given at the beginning of this section will be said to be differentiable 'as complex functions'. (After this section, the word 'differentiable' will always be used to mean 'differentiable as a complex function'.) Our next task is to compare the two definitions.

Let A be a subset of $\mathbf{R}^2$, and let u be a function from A to $\mathbf{R}$. Let (a,b) be an interior point of A. Then u is said to be **differentiable** (as a function on $\mathbf{R}^2$) at (a,b) if there exist real numbers λ, μ such that, given $\epsilon > 0$, there exists $\delta > 0$ such that, for all real h, k with $\sqrt{(h^2+k^2)} < \delta$,

$$|u(a+h,b+k)-u(a,b)-(\lambda h+\mu k)| \leqslant \epsilon\sqrt{(h^2+k^2)}.$$

Keeping k equal to 0, we see that

$$\lambda = \lim_{h\to 0}\frac{1}{h}(u(a+h,b)-u(a,b)).$$

In other words, λ is the first partial derivative of f at (a,b); we denote this partial derivative by $D_1 u(a,b)$. Similarly, μ is the second partial derivative $D_2 u(a,b)$.

A function f from A to $\mathbf{R}^2$ can be expressed as a pair of functions u, v from A to $\mathbf{R}$ by writing

$$f(x,y) = (u(x,y), v(x,y)),$$

and we say that f is differentiable (as a function on $\mathbf{R}^2$) at (a,b) if u and v are. The reader may be acquainted with the concept of a differentiable function from $\mathbf{R}^n$ to $\mathbf{R}^m$, or, more generally, from one normed linear space to another. However, familiarity with this concept is not required here.

1.5.11 Theorem. *Let f be a function from A to $\mathbf{C}$, where $A \subseteq \mathbf{C}$, and let $a+ib$ be an interior point of A. Write*

$$u(x,y) = \operatorname{Re} f(x+iy), \qquad v(x,y) = \operatorname{Im} f(x+iy).$$

Then f is differentiable (as a complex function) at $a+ib$ if and only if the following conditions hold:

(i) *u and v are differentiable (as functions on $\mathbf{R}^2$) at (a,b),*
and (ii) $D_1 u(a,b) = D_2 v(a,b)$, $D_2 u(a,b) = -D_1 v(a,b)$.

When this is true, we have

$$\begin{aligned} f'(a+ib) &= D_1 u(a,b) + iD_1 v(a,b) \\ &= D_2 v(a,b) - iD_2 u(a,b). \end{aligned}$$

Proof. Suppose that $f'(a+ib) = \lambda + i\mu$. Then, given $\epsilon > 0$, there exists $\delta > 0$ such that for all real h, k with $|h+ik| < \delta$,

$$|f((a+h)+i(b+k)) - f(a+ib) - (\lambda + i\mu)(h+ik)| \leqslant \epsilon |h+ik|.$$

Taking the real and imaginary parts, we obtain

$$\left.\begin{aligned} |u(a+h,b+k) - u(a,b) - (\lambda h - \mu k)| &\leqslant \epsilon\sqrt{(h^2+k^2)}, \\ |v(a+h,b+k) - v(a,b) - (\mu h + \lambda k)| &\leqslant \epsilon\sqrt{(h^2+k^2)}. \end{aligned}\right\} \qquad (1)$$

Hence u and v are differentiable (as functions on $\mathbf{R}^2$) at (a,b), and

$$\begin{aligned} D_1 u(a,b) &= D_2 v(a,b) = \lambda, \\ -D_2 u(a,b) &= D_1 v(a,b) = \mu. \end{aligned}$$

Conversely, suppose that conditions (i) and (ii) hold, and write $\lambda = D_1 u(a,b) = D_2 v(a,b)$, $\mu = -D_2 u(a,b) = D_1 v(a,b)$. Given $\epsilon > 0$, there exists $\delta > 0$ such that the inequalities (1) hold for all real h, k with $\sqrt{(h^2+k^2)} \leqslant \delta$. For such h, k, we have

$$|f((a+h)+i(b+k)) - f(a+ib) - (\lambda + i\mu)(h+ik)| \leqslant 2\epsilon|h+ik|,$$

showing that $f'(a+ib)$ exists and equals $\lambda + i\mu$.

The equalities in condition (ii) of 1.5.11 are called the 'Cauchy-Riemann equations'. One striking consequence of these equations and an earlier result is the following:

1.5.12 Corollary. *Let G be a connected, open subset of $\mathbf{C}$. If f is differentiable as a complex function, and is real-valued on G, then f is constant on G.*

Proof. With the above notation, $v = 0$, and hence $f'(z) = 0$ $(z \in G)$. The statement now follows from 1.5.10.

A sufficient condition for u to be differentiable (as a function on $\mathbf{R}^2$) at (a,b) is that $D_1 u$ and $D_2 u$ should exist in a neighbourhood of (a,b) and be continuous at (a,b) (the reader may either treat this statement as an exercise or refer to a book on the subject, for example Spivak [7], page 31). Together with 1.5.11, this yields:

1.5.13 Corollary. *Let u, v be functions from $\mathbf{R}^2$ to $\mathbf{R}$ such that the partial derivatives of u and v exist in a neighbourhood of (a,b) and* (i) *are continuous at (a,b), and* (ii) *satisfy the Cauchy-Riemann equations at (a,b). Then $u+iv$ is differentiable (as a complex function) at $a+ib$.*

Example. To find the set of points in $\mathbf{C}$ at which f is differentiable as a complex function, where $f(x+iy) = x^2 + iy^2$.

With the notation of 1.5.11, $u(x,y) = x^2$, $v(x,y) = y^2$, so

$$D_1 u(x,y) = 2x, \qquad D_2 u(x,y) = 0,$$
$$D_1 v(x,y) = 0, \qquad D_2 v(x,y) = 2y.$$

These partial derivatives are continuous everywhere, and satisfy the Cauchy-Riemann equations if and only if $x = y$. Hence f is differentiable (as a complex function) at the points $(1+i)x$ $(x \in \mathbf{R})$.

Reverting to our original question on the comparison between the two definitions of differentiability, we notice that 1.5.11 can be stated very economically as follows: f is differentiable as a complex function if and only if it is differentiable as a function on $\mathbf{R}^2$ and the matrix of its partial derivatives has the form

$$\begin{pmatrix} \lambda & -\mu \\ \mu & \lambda \end{pmatrix}$$

The reader can easily verify that if f is defined by $f(x+iy) = x$, then f is differentiable everywhere as a function on $\mathbf{R}^2$, but differentiable nowhere as a complex function.

Exercises 1.5

1 Prove directly from the definition (i.e. without using the Cauchy-Riemann equations) that $z \mapsto \bar{z}$ is differentiable nowhere (as a complex function), and that $z \mapsto z\bar{z}$ is differentiable only at 0.

2 Find the sets of points at which the following functions are differentiable (as complex functions):
(i) $f(x+iy) = x^2 + 2ixy$,
(ii) $f(x+iy) = 2xy + i(x + \frac{2}{3}y^3)$.

3 Show that the only differentiable complex functions of the form $f(x+iy) = u(x) + iv(y)$ (where u and v are real functions) are given by $f(z) = \lambda z + c$, where $\lambda \in \mathbf{R}$ and $c \in \mathbf{C}$.

4 If f is differentiable (as a complex function) and $|f|$ is constant on $D(a,r)$, show that f is constant on this disc.

5 Define $f(0) = 0$ and

$$f(x+iy) = \frac{(1+i)\,x^3 - (1-i)\,y^3}{x^2+y^2} \qquad (x+iy \neq 0).$$

Show that f satisfies the Cauchy-Riemann equations, but is not differentiable, at 0. (Note: there is no need to compute the partial derivatives at points other than 0.)

6 Show that if λ is real and $\zeta = \lambda + i(1-\lambda)$, then $|\zeta|^2 > \frac{1}{2}$. Deduce that if $f(z) = z^3$, then there is no point ζ on the line segment $[1:i]$ such that $f(i) - f(1) = (i-1)f'(\zeta)$.

7 Let u be the real part of a differentiable complex function defined on an open subset A of $\mathbf{C}$. Suppose that the second-order partial derivatives of u exist and are continuous on A. Prove that $D_{1,1}u + D_{2,2}u = 0$ on A (i.e. that u is *harmonic* on A).

8 Use 1.5.11 to give an alternative proof of 1.5.10. (Hint: modify 1.4.6.)

1.6. The exponential and trigonometric functions

By providing useful counter-examples, and by making it possible to define a^b, the exponential and trigonometric functions play an important part in real analysis. In complex analysis, these special functions are of even greater importance. They enable us to express every complex number in the very useful form re^{it} (from which, we can deduce, for example, the existence of nth roots). They also make it possible to turn the Euclidean definition of a circle (the set of points at a certain distance from a given point) into a path; virtually all of our later theory is dependent on the use of circular paths. Other uses of these functions will become apparent later (see especially 3.2).

Starting from the power-series definitions, we shall deduce all the properties of these functions that are required in this book. It is perhaps worth pointing out that in doing so, we have to make use of most of the standard theorems of elementary real and complex analysis. Our development is self-contained, and does not assume previous knowledge of the real exponential and trigonometric functions. Two alternative proofs (one using differentiation and one avoiding it) are given in a few places where this seems illuminating.

For all z in $\mathbf{C}$, define:

$$\exp z = e^z = \sum_{n=0}^{\infty} \frac{z^n}{n!},$$

$$\cos z = \sum_{n=0}^{\infty} (-1)^n \frac{z^{2n}}{(2n)!} = 1 - \frac{z^2}{2!} + \frac{z^4}{4!} - \cdots,$$

$$\sin z = \sum_{n=0}^{\infty} (-1)^n \frac{z^{2n+1}}{(2n+1)!} = z - \frac{z^3}{3!} + \frac{z^5}{5!} - \cdots.$$

Each series converges for all complex z, by D'Alembert's ratio test, or by comparison with $\sum (|z|/k)^n$, where $k > |z|$. We use the notations $\exp z$ and e^z interchangeably, depending on typographical convenience. The following properties are immediate consequences of the definitions:

$$\cos z = \tfrac{1}{2}(e^{iz} + e^{-iz}), \qquad \sin z = \frac{1}{2i}(e^{iz} - e^{-iz}),$$

$$e^{iz} = \cos z + i \sin z, \qquad e^{-iz} = \cos z - i \sin z,$$

$$\cos(-z) = \cos z, \qquad \sin(-z) = -\sin z,$$

$$e^0 = 1, \qquad \cos 0 = 1, \qquad \sin 0 = 0,$$

$$\lim_{z \to 0} \frac{\sin z}{z} = 1 \qquad \text{(by 1.2.7)}.$$

The number e^1 is denoted by e. We use the usual notation for ratios of trigonometric functions. For example, we define $\tan z$ to be $\sin z / \cos z$ whenever $\cos z \neq 0$.

By 1.5.7, we have:

1.6.1. $\dfrac{d}{dz} e^z = e^z, \quad \dfrac{d}{dz} \cos z = -\sin z, \quad \dfrac{d}{dz} \sin z = \cos z.$

The importance of exp depends largely on its additive property, which we now establish.

1.6.2 Theorem. $e^{a+b} = e^a e^b$ *for all complex* a, b.

Proof 1. Since the defining series for e^a and e^b are absolutely convergent, 1.2.2 applies to show that $e^a e^b = \sum_{n=0}^{\infty} c_n$, where

$$\begin{aligned} c_n &= \sum_{k=0}^{n} \frac{a^k}{k!} \frac{b^{n-k}}{(n-k)!} \\ &= \frac{1}{n!} \sum_{k=0}^{n} \binom{n}{k} a^k b^{n-k} \\ &= \frac{1}{n!} (a+b)^n. \end{aligned}$$

Proof 2. Choose a in $\mathbf{C}$, and let $f(z) = e^z e^{a-z}$ $(z \in \mathbf{C})$. By 1.6.1 and 1.5.2, $f'(z) = 0$ $(z \in \mathbf{C})$. Therefore, by 1.5.10, f is constant on $\mathbf{C}$. Since $f(0) = e^a$, this shows that $e^z e^{a-z} = e^a$ for all a, z in $\mathbf{C}$, which is equivalent to the required statement.

1.6.3 Corollary. *For all z in $\mathbf{C}$, $e^z \neq 0$ and $e^{-z} = 1/e^z$.*

1.6.4 Corollary. *For all z in $\mathbf{C}$, $\cos^2 z + \sin^2 z = 1$.*

Proof.

$$\begin{aligned} \cos^2 z + \sin^2 z &= (\cos z + i \sin z)(\cos z - i \sin z) \\ &= e^{iz} e^{-iz} = 1. \end{aligned}$$

1.6.5 Corollary. *For all x in $\mathbf{R}$, $|e^{ix}| = 1$ and $\cos x$, $\sin x \in [-1, 1]$.*

1.6.6 Corollary. *The function* exp *is strictly positive and strictly increasing on $\mathbf{R}$. $e^x \to \infty$ and $e^{-x} \to 0$ as $x \to \infty$. Further, if n is a positive integer, then $x^{-n} e^x \to \infty$ and $x^n e^{-x} \to 0$ as $x \to \infty$.*

Proof. It is immediate from the series that e^x is strictly positive and strictly increasing for $x > 0$, and that

$$x^{-n} e^x > \frac{x}{(n+1)!},$$

which tends to ∞ as $x \to \infty$. Since $e^{-x} = 1/e^x$, the other statements follow.

Notice that the last two corollaries describe the behaviour of exp along the imaginary and real axes respectively.

Next, we derive the addition formulae for cos and sin. For real a and b, these can be obtained by equating the real and imaginary parts in the identity

$$e^{i(a+b)} = e^{ia}e^{ib} = (\cos a + i\sin a)(\cos b + i\sin b).$$

We now show that the formulae are also valid for complex a, b:

1.6.7. *For all* a, b *in* $\mathbf{C}$,

$$\cos(a+b) = \cos a\cos b - \sin a\sin b,$$

$$\sin(a+b) = \sin a\cos b + \cos a\sin b.$$

Proof 1. $\cos a\cos b - \sin a\sin b$

$$= \tfrac{1}{4}(e^{ia}+e^{-ia})(e^{ib}+e^{-ib}) + \tfrac{1}{4}(e^{ia}-e^{-ia})(e^{ib}-e^{-ib})$$

$$= \tfrac{1}{2}(e^{ia+ib}+e^{-ia-ib})$$

$$= \cos(a+b).$$

Proof 2. Choose a in $\mathbf{C}$, and let

$$f(z) = \cos z\cos(a-z) - \sin z\sin(a-z) \qquad (z \in \mathbf{C}).$$

On writing out the terms, we find that $f'(z) = 0$ $(z \in \mathbf{C})$. Hence $f(z) = f(0) = \cos a$ $(z \in \mathbf{C})$.

In both cases, a similar proof applies to sin.

Next we mention some simple facts concerning the moduli of exp, cos and sin. If $z = x+iy$, then $e^z = e^x e^{iy}$, so, by 1.6.5, $|e^z| = e^x$.

It is customary to define:

$$\cosh z = \tfrac{1}{2}(e^z + e^{-z}) = 1 + \frac{z^2}{2!} + \frac{z^4}{4!} + \cdots,$$

$$\sinh z = \tfrac{1}{2}(e^z - e^{-z}) = z + \frac{z^3}{3!} + \frac{z^5}{5!} + \cdots.$$

Then $\cosh z = \cos iz$ and $i\sinh z = \sin iz$, so, regarded as complex functions, cosh and sinh are simply compositions of cos and sin with trivial functions, and have all the same properties. Regarded as real functions, on the other hand, cos and cosh look quite different, being in fact the values of the same complex function along two perpendicular axes. In particular, it is clear from the series that cosh and sinh are strictly positive and strictly increasing for $x > 0$, and that both tend to ∞ as $x \to \infty$. Their relevance to the moduli of cos and sin is apparent from the next result.

1.6.8. *If $x, y \in \mathbf{R}$ and $y > 0$, then*

$$\sinh y \leqslant \left| \begin{matrix}\cos \\ \sin\end{matrix} (x \pm iy) \right| \leqslant \cosh y.$$

Proof. We have $\cos(x+iy) = \frac{1}{2}(e^{-y+ix} + e^{y-ix})$. Since $e^y > 1 > e^{-y}$ for $y > 0$, it follows that

$$\tfrac{1}{2}(e^y - e^{-y}) \leqslant \cos(x+iy) \leqslant \tfrac{1}{2}(e^y + e^{-y}).$$

Similar reasoning applies in the other cases.

Observe that the real and imaginary parts of cos and sin are given by:

$$\cos(x+iy) = \cos x \cosh y - i \sin x \sinh y,$$

$$\sin(x+iy) = \sin x \cosh y + i \cos x \sinh y.$$

Periodicity and zeros

1.6.9. *There is a unique positive number π such that* $\cos(\pi/2) = 0$, $\sin(\pi/2) = 1$, *and* cos *is strictly decreasing,* sin *strictly increasing, on* $[0, \pi/2]$.

Proof. We make repeated use of the fact (proved by the mean-value theorem or by integration) that if f is a differentiable real function and $f'(t) \geqslant 0$ for $0 \leqslant t \leqslant x$, then $f(x) \geqslant f(0)$.

For all $x \geqslant 0$, $1 - \cos x \geqslant 0$, by 1.6.5. This is the derivative of $x - \sin x$, so $x - \sin x \geqslant 0$ $(x \geqslant 0)$. This is the derivative of $\frac{1}{2}x^2 + \cos x$, so $\frac{1}{2}x^2 + \cos x \geqslant \cos 0 = 1$ $(x \geqslant 0)$. Repeating the process twice more, we obtain

$$\sin x \geqslant x - \tfrac{1}{6}x^3 \qquad (x \geqslant 0) \qquad (1),$$

$$\cos x \leqslant 1 - \tfrac{1}{2}x^2 + \tfrac{1}{24}x^4 \qquad (x \geqslant 0) \qquad (2).$$

In particular, $\cos(\sqrt{3}) \leqslant 1 - \frac{3}{2} + \frac{9}{24} = -\frac{1}{8}$. Since cos is continuous, the intermediate value theorem shows that there is a number $\pi/2$ such that $0 < \pi/2 < \sqrt{3}$ and $\cos(\pi/2) = 0$. If $0 < x \leqslant \sqrt{3}$, then

$$\sin x \geqslant x(1 - \tfrac{1}{6}x^2) \geqslant \tfrac{1}{2}x > 0,$$

by (1). By the mean value theorem, it follows that cos is strictly decreasing on $[0, \pi/2]$, and therefore strictly positive on $[0, \pi/2)$. Again by the mean value theorem, it now follows that sin is strictly increasing on $[0, \pi/2]$. By 1.6.4, $\sin^2(\pi/2) = 1$. Since $\sin(\pi/2) > 0$, it follows that $\sin(\pi/2) = 1$.

Hence we have $e^{\pi i/2} = i$, $e^{\pi i} = -1$, $e^{n\pi i} = (-1)^n$, $\cos n\pi = (-1)^n$, $\sin n\pi = 0$ (where n is an integer). Also, $e^{z+2\pi i} = e^z$ for all z, and the addition formulae for cos and sin give:

$$\cos(z + 2\pi) = \cos z, \qquad \sin(z + 2\pi) = \sin z,$$

$$\cos\left(\frac{\pi}{2} - z\right) = \sin z, \qquad \sin\left(\frac{\pi}{2} - z\right) = \cos z,$$

for all z in $\mathbf{C}$. It is now a simple matter to find all the points where cos or sin takes the value 0, and all the points where exp takes the value 1 (we have seen already that exp is never zero).

1.6.10. $\sin z = 0$ *if and only if* $z = n\pi$ *for some integer* n. $\cos z = 0$ *if and only if* $z = (n + \frac{1}{2})\pi$ *for some integer* n.

Proof. By 1.6.8, $\sin z \neq 0$ if $z \notin \mathbf{R}$. Now $\sin x \neq 0$ for $0 < x \leqslant \pi/2$, by 1.6.9. Since $\sin[(\pi/2) + x] = \sin[(\pi/2) - x] = \cos x$, it follows that $\sin x \neq 0$ for $0 < x < \pi$. Since $\sin(-x) = -\sin x$, the only zeros of sin in $[-\pi, \pi]$ are $-\pi, 0, \pi$. The first statement now follows, using the fact that $\sin(x + 2\pi) = \sin x$. The identity $\cos z = \sin[z + (\pi/2)]$ then gives the second statement.

1.6.11. $e^z = 1$ *if and only if* $z = 2n\pi i$ *for some integer* n.

Proof. Suppose that $e^z = 1$, where $z = x + iy$. Then $|e^z| = e^x = 1$, so $x = 0$. Hence $\cos y = 1$, $\sin y = 0$. By 1.6.10, this occurs only for $y = 2n\pi$ (n an integer).

Expression of a complex number in polar form

1.6.12 Theorem. *For any non-zero complex number z, there is a unique number t in $[0,2\pi)$ such that $z = |z|\, e^{it}$.*

Proof. If t, $u \in [0,2\pi)$ and $z = |z|\, e^{it} = |z|\, e^{iu}$, then $e^{i(t-u)} = 1$, so $t = u$, by 1.6.11.

We prove the existence of t by considering the four quadrants separately. Let $z/|z| = x + iy$, so that $x^2 + y^2 = 1$.

(i) $x \geqslant 0$, $y \geqslant 0$. By the intermediate value theorem, there exists t in $[0,\pi/2]$ such that $\cos t = x$. Then $\sin^2 t = y^2$ and $\sin t \geqslant 0$, so $\sin t = y$. This gives $z = |z|\, e^{it}$.

(ii) $x \leqslant 0$, $y \geqslant 0$. Then $x + iy = i(y - ix)$. By (i), there exists t in $[0,\pi/2]$ such that $y - ix = e^{it}$. Then $x + iy = e^{i[t+(\pi/2)]}$.

(iii) $x \leqslant 0$, $y \leqslant 0$. By (i), there exists t in $[0,\pi/2]$ such that $-x - iy = e^{it}$. Then $x + iy = e^{i(t+\pi)}$.

(iv) $x \geqslant 0$, $y \leqslant 0$. Then $x + iy = -i(-y + ix)$. By (i), there exists t in $[0,\pi/2]$ such that $-y + ix = e^{it}$. Then $x + iy = e^{i[t+(3\pi/2)]}$.

If z is a non-zero complex number, then any real number t such that $z = |z|\, e^{it}$ is said to be an **argument** of z. The arguments of z are therefore $\{t_0 + 2n\pi : n = 0, \pm 1, \pm 2, \ldots\}$ for a unique t_0 in $[0,2\pi)$.

Notice that if $z = re^{it}$, then $\bar{z} = re^{-it}$ and $1/z = 1/re^{-it}$.

Circles and angles

The Euclidean definition of a circle is the set of all points at a certain distance from a given point. Now by 1.6.12, the set of all points in $\mathbf{C}$ whose distance from a is r is

$$\{a + re^{it} : 0 \leqslant t \leqslant 2\pi\}.$$

Hence we have a ready-made definition of this set as the range of a smooth closed path: the path $t \mapsto a + re^{it}$ $(0 \leqslant t \leqslant 2\pi)$ will be called the **circle centre** a, **radius** r, and will be denoted by $C(a,r)$. The circle centre 0, radius 1 is called the **unit circle**. By 1.4.7, the length of the arc $t \mapsto a + re^{it}$ $(\alpha \leqslant t \leqslant \beta)$ is

$$\int_{\alpha}^{\beta} |ir\, e^{it}|\, dt = (\beta - \alpha)\, r,$$

and the length of the whole circle is $2\pi r$.

The point z is said to be **inside** $C(a,r)$ if $|z-a| < r$, and **outside** $C(a,r)$ if $|z-a| > r$.

A subset of $\mathbf{C}$ bounded by two half-lines meeting at z_0 is necessarily of the form

$$\{z_0 + r e^{it} : \alpha \leqslant t \leqslant \beta, r \geqslant 0\}$$

where $\alpha < \beta < \alpha + 2\pi$. We can define the associated **angle** to be $\beta - \alpha$. This definition is quite superfluous to our purposes; we mention it only to emphasize the fact that in order to give any meaning to the notion of 'angle' (so frequently used without definition), it is necessary to use the exponential function.

nth roots

We proved in 1.1 that every complex number has square roots. The polar form enables us to prove (by a much neater and more instructive method) that every complex number has nth roots for each positive integer n. First we consider roots of unity: suppose that $z^n = 1$. Then $|z| = 1$, so $z = e^{it}$ for some t in $[0, 2\pi)$, by 1.6.12. By 1.6.11, $nt = 2r\pi$ for some integer r. Hence there are exactly n distinct nth roots of 1 in $\mathbf{C}$, viz. $\exp(2r\pi i/n)$ $(r = 0, 1, \ldots, n-1)$, or $1, \omega, \ldots, \omega^{n-1}$, where $\omega = \exp(2\pi i/n)$.

1.6.13. *Any non-zero complex number has exactly n distinct nth roots.*

Proof. Let $a = |a| e^{it}$. Then $z^n = a$, where $z = |z|^{1/n} \exp(it/n)$. If $z_1^n = a$, then $z_1 = wz$, where $w^n = 1$. Hence the distinct nth roots of a are z, ωz, ..., $\omega^{n-1} z$, where $\omega = \exp(2\pi i/n)$.

Continuous choice of argument

We now ask to what extent it is possible to choose an argument of each non-zero complex number so as to form a continuous function. The answer is that this cannot be done on the whole of $\mathbf{C} \backslash \{0\}$, but that it can be done on the 'cut plane' consisting of $\mathbf{C}$ with one half-line removed. For each real α, define

$$H_\alpha = \{-r e^{i\alpha} : r \geqslant 0\}.$$

Loosely speaking, H_α is the half-line starting at 0 and pointing away from $e^{i\alpha}$.

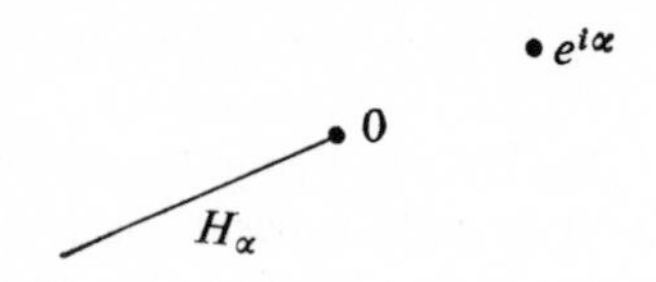

1.6.14 Lemma. $\mathbf{C}\backslash H_\alpha$ *is open and star-shaped about* $e^{i\alpha}$.

Proof. Write $e^{i\alpha} = c$. Suppose that a sequence $\{-r_n c\}$ of points of H_α is convergent. Then $\{r_n\}$ is convergent to some $r \geqslant 0$, so $\lim_{n\to\infty}(-r_n c) = -rc$, which is in H_α. Hence H_α is closed.

To show that $\mathbf{C}\backslash H_\alpha$ is star-shaped about c, take d in $\mathbf{C}$ and suppose that, for some t in $(0, 1]$, $(1-t)c + td$ is in H_α, so is of the form $-rc$, where $r \geqslant 0$. Then $d = -(1/t)(1-t+r)c$, which is in H_α. Hence if d is in $\mathbf{C}\backslash H_\alpha$, then so is every point of $[c:d]$.

For z in $\mathbf{C}\backslash H_\alpha$, we define $\arg_\alpha(z)$ to be the unique argument of z in $(\alpha - \pi, \alpha + \pi)$.

1.6.15. $\arg_\alpha$ *is continuous on* $\mathbf{C}\backslash H_\alpha$. *If* θ *is a continuous function on* $\mathbf{C}\backslash H_\alpha$ *such that, for each* z, $\theta(z)$ *is an argument of* z, *then there is an integer* k *such that* $\theta(z) = \arg_\alpha(z) + 2k\pi$ *for all* z *in* $\mathbf{C}\backslash H_\alpha$.

Proof. Let D be the set of z in $\mathbf{C}\backslash H_\alpha$ with $|z| = 1$. For z in D, let $g(z) = \arg_\alpha(z)$. We show that g is continuous on D. Since $\arg_\alpha(z) = g(z/|z|)$ $(z \in \mathbf{C}\backslash H_\alpha)$ and $z \mapsto z/|z|$ is continuous, it will follow that $\arg_\alpha$ is continuous on $\mathbf{C}\backslash H_\alpha$.

Take t_0 in $(\alpha - \pi, \alpha + \pi)$, and ϵ such that

$$\alpha - \pi < t_0 - \epsilon < t_0 + \epsilon < \alpha + \pi.$$

Let

$$E = \{e^{it} : \alpha - \pi \leqslant t \leqslant t_0 - \epsilon\} \cup \{e^{it} : t_0 + \epsilon \leqslant t \leqslant \alpha + \pi\}.$$

Since closed real intervals are compact and the mapping $t \mapsto e^{it}$ is continuous, 0.6 shows that E is compact, and therefore closed.

But $e^{it_0} \notin E$, so there exists $\delta > 0$ such that if $|z| = 1$ and $|z - e^{it_0}| < \delta$, then $z \notin E$. For such z, we must have $z = e^{iu}$ for some u in $(t_0 - \epsilon, t_0 + \epsilon)$. Clearly, $u = g(z)$. We have shown that if $z \in D$ and $|z - e^{it_0}| < \delta$, then $|g(z) - t_0| < \epsilon$, in other words, that g is continuous at e^{it_0}.

Finally, suppose that the function θ is as stated. Then for each z in $\mathbf{C}\backslash H_\alpha$, there is an integer $k(z)$ such that $\arg_\alpha(z) - \theta(z) = 2k(z)\pi$. Now $\mathbf{C}\backslash H_\alpha$ is star-shaped, and so certainly connected. Since k is continuous on $\mathbf{C}\backslash H_\alpha$, it follows that it is constant there.

For alternative proofs of this theorem, see Duncan [2], page 59, and Moss–Roberts [5], page 158.

It is now clear that it is impossible to choose arguments so as to define a continuous function on the whole of $\mathbf{C}\backslash\{0\}$. For if θ were the supposed continuous function, then there would be an integer k such that $\theta = \arg_\alpha + 2k\pi$ on $\mathbf{C}\backslash H_\alpha$. But it is evident that no choice of arguments along H_α will make $\arg_\alpha$ continuous there.

Logarithms

Any complex number w such that $e^w = z$ is said to be a **logarithm** of z. It follows from 1.6.6 that each strictly positive real number x has a unique real logarithm, which we denote by $\log x$. The mapping $x \mapsto \log x$ $(x > 0)$ is continuous, since it is the inverse of an increasing continuous function. Clearly, the logarithms of z are the numbers $\log|z| + it$, where t is any argument of z (and z is non-zero).

If α is real, we define

$$\log_\alpha(z) = \log|z| + i\arg_\alpha(z)$$

for z in $\mathbf{C}\backslash H_\alpha$. By 1.6.15 and the fact that $z \mapsto \log|z|$ is continuous, the function $\log_\alpha$ is continuous on $\mathbf{C}\backslash H_\alpha$. But it is the inverse of exp on

$$\{x + iy : \alpha - \pi < y < \alpha + \pi\},$$

so we can use the rule for differentiation of inverse functions (1.5.5) to prove:

1.6.16. $\dfrac{d}{dz}\log_\alpha(z) = \dfrac{1}{z}$ *for* z *in* $\mathbf{C}\backslash H_\alpha$.

Proof. Write $g = \log_\alpha$, and $f(z) = e^z$ ($|\text{Im}\, z - \alpha| < \pi$). Suppose that $f(a) = b$. By 1.5.5,

$$g'(b) = \frac{1}{f'(a)} = \frac{1}{e^a} = \frac{1}{b}.$$

In the real case, it follows from the 'fundamental theorem of calculus' that $\log x = \int_1^x (1/t)dt$. When we have defined integration of complex functions, we shall see that an analogous statement is true in the complex case.

1.6.17. *If* $|z| < 1$, *then* $\log_0(1+z) = \sum_{n=1}^{\infty} (-1)^{n-1} z^n/n$.

Proof. Let $f(z) = \log_0(1+z)$, $g(z) = \sum_{n=1}^{\infty} (-1)^{n-1} z^n/n$ ($|z| < 1$). Then $f'(z) = 1/(1+z)$, by 1.6.16, and

$$g'(z) = \sum_{n=0}^{\infty} (-z)^n = \frac{1}{1+z},$$

by 1.5.7. Therefore by 1.5.10, $f - g$ is constant on $D(0,1)$. Since $f(0) = g(0) = 0$, the result follows.

By analogy with real numbers, it is natural to want to define z^λ to be $\exp(\lambda \log z)$ when $\lambda \in \mathbf{C}$ and $z \in \mathbf{C}\backslash\{0\}$. However, the result will clearly depend on which logarithm of z we choose. We can, in fact, define a whole family of functions (each continuous on a 'cut plane') each of which might claim to be z^λ. For each real α, define

$$p_\alpha^\lambda(z) = \exp(\lambda \log_\alpha(z))$$

for z in $\mathbf{C}\backslash H_\alpha$. Then p_α^λ is differentiable on its domain, and $p_\alpha^\lambda(z) p_\alpha^\mu(z) = p_\alpha^{\lambda+\mu}(z)$. The function p_0^λ has an especially strong claim, in the light of the following facts:

(i) If $x > 0$ and λ is real, then $p_0^\lambda(x) = x^\lambda$;

(ii) for all λ in $\mathbf{C}$, $p_0^\lambda(e) = e^\lambda$.

Notice that, for fixed z and α, the mapping $\lambda \mapsto p_\alpha^\lambda(z)$ is defined and differentiable for all λ in $\mathbf{C}$.

The evaluation of π

Given the power-series definition of cos, it might seem a formidable problem to find a series that converges to the first positive zero of cos (the existence of which has been proved above). The following solution to this problem is another pleasant application of the theorems of elementary real analysis.

By 1.6.9, the function tan is strictly increasing on $(-\pi/2, \pi/2)$, and takes all real values on this range. Let arctan be the inverse function. From 1.6.1, we have $d/dx \tan x = 1 + \tan^2 x$ $[-\pi/2 < x < \pi/2]$, so, by the rule for differentiating an inverse function,

$$\frac{d}{dx} \arctan x = \frac{1}{1+x^2} \qquad (x \in \mathbf{R}).$$

Therefore, by the fundamental theorem of calculus,

$$\arctan x = \int_0^x \frac{1}{1+t^2}\, dt \qquad (x \in \mathbf{R}).$$

1.6.18. *For* $-1 \leqslant x \leqslant 1$,

$$\arctan x = \sum_{n=0}^{\infty} (-1)^n \frac{x^{2n+1}}{2n+1}.$$

Proof. Let $s_n(t) = 1 - t^2 + t^4 - \cdots + (-1)^n t^{2n}$. Then

$$(1+t^2)\, s_n(t) = 1 + (-1)^n t^{2n+2}.$$

Hence

$$\arctan x = \int_0^x s_n + r_n(x),$$

where

$$r_n(x) = (-1)^{n+1} \int_0^x \frac{t^{2n+2}}{1+t^2}\, dt.$$

If $|x| \leqslant 1$, then

$$|r_n(x)| \leqslant \left| \int_0^x t^{2n+2}\, dt \right| = \frac{|x|^{2n+3}}{2n+3},$$

so $r_n(x) \to 0$ as $n \to \infty$. The result follows.

Since $\cos 2x = \cos^2 x - \sin^2 x$, we have $\cos(\pi/4) = \sin(\pi/4)$, so $\tan(\pi/4) = 1$, and

$$\frac{\pi}{4} = \arctan 1 = 1 - \tfrac{1}{3} + \tfrac{1}{5} - \cdots + \frac{(-1)^n}{2n+1} + \cdots.$$

To evaluate π in practice, it is better to use series that converge faster through having x smaller. By 1.6.7, if $\cos x$, $\cos y \neq 0$ and $\tan x \tan y \neq 1$, then

$$\tan(x+y) = \frac{\tan x + \tan y}{1 - \tan x \tan y}.$$

In particular, if $\tan x = \frac{1}{2}$ and $\tan y = \frac{1}{3}$, then $\tan(x+y) = 1$. Hence

$$\begin{aligned}\frac{\pi}{4} &= \arctan\tfrac{1}{2} + \arctan\tfrac{1}{3}\\ &= \sum_{n=0}^{\infty} \frac{(-1)^n}{2n+1}\left(\frac{1}{2^{2n+1}} + \frac{1}{3^{2n+1}}\right).\end{aligned}$$

Even more rapid convergence is given by

$$\frac{\pi}{4} = 4\arctan\tfrac{1}{5} - \arctan\tfrac{1}{239}.$$

Exercises 1.6

1 What are the zeros of $1+e^z$, $\sinh z$, $\cosh z$, $1/e - e^z$, $1+i-e^z$?

2 If t and u are real, show that $\frac{1}{2}(t+u)$ is an argument of $e^{it}+e^{iu}$, and find the modulus. (Note: this question has a one-line solution.)

3 Prove that $(1-\cos x)/x \to 0$ as $x \to 0$, and deduce that $n(\omega_n - 1) \to 2\pi i$ as $n \to \infty$, where $\omega_n = \exp(2\pi i/n)$.

4 If x is real and not an integer multiple of 2π, prove that

$$1 + \cos x + \cdots + \cos(n-1)x = \cos\tfrac{1}{2}(n-1)x \frac{\sin(n/2)x}{\sin\frac{1}{2}x},$$

$$\sin x + \cdots + \sin(n-1)x = \sin\tfrac{1}{2}(n-1)x \frac{\sin(n/2)x}{\sin\frac{1}{2}x}.$$

5 If $y \neq 0$, show that $\cos(x+iy)$ is real if and only if x is an integer multiple of π. Write down the set of z for which $\sin z$ is real, and deduce that if both $\cos z$ and $\sin z$ are real, then z is real.

6 Let $f(z) = z\sin(1/z)$ $(z \neq 0)$, $f(0) = 0$. Is f continuous at 0?

7 If f is differentiable and non-zero at z_0, and n is a positive integer, show that there is a neighbourhood B of z_0 and a differentiable function h on B such that $f(z) = h(z)^n$ $(z \in B)$. (Hint: $|f(z) - f(z_0)| < |f(z_0)|$ for z in a neighbourhood of z_0).

8 For any z in $\mathbf{C}$, prove that $n\log_0(1 + z/n)$ is defined for all sufficiently large n and tends to z as $n \to \infty$. Deduce that $(1 + z/n)^n$ tends to e^z as $n \to \infty$.

9 Show that tan gives a one-to-one mapping of

$$\left\{z: -\frac{\pi}{2} < \operatorname{Re} z \leqslant \frac{\pi}{2}\right\} \Big\backslash \left\{\frac{\pi}{2}\right\}$$

on to $\mathbf{C}\backslash\{i, -i\}$.

10 Let f be a continuous real-valued function on a real interval $[a, b]$. For each z in $\mathbf{C}$, define

$$F(z) = \int_a^b e^{-zt} f(t)\, dt.$$

By writing $e^z = 1 + z + z^2 g(z)$, or otherwise, show that

$$F'(z) = -\int_a^b t e^{-zt} f(t)\, dt$$

for all z. (F is said to be the *Laplace transform* of f.)

11 Let f be a continuous, real-valued function on $\{z: |z| = 1\}$. Prove that there exists a z such that $f(-z) = f(z)$.

1.7. Integration

The Riemann integral of a real function is defined on an interval. For a complex function, it is defined on a rectifiable path. The general idea is as follows. Let φ be a path, with domain $[c, d]$, and let f be a complex-valued function defined at least on φ^*. Let $c = t_0 < t_1 < \cdots < t_n = d$ be a dissection of $[c, d]$, and suppose that $f(\varphi(t))$ is close to λ_j for $t_{j-1} < t < t_j$. Then $\int_\varphi f$ is to be a number approximated by the sum

$$\sum_{j=1}^{n} \lambda_j[\varphi(t_j) - \varphi(t_{j-1})] \tag{1}$$

This sum does not express an area or volume – it is simply written down by algebraic analogy with the real case. We observe that it makes use of complex multiplication.

The λ_j above are approximations to the values of $f \circ \varphi$, which is a complex-valued function on $[c, d]$. Our problem can therefore be rephrased in the following terms. Let h be a complex-valued function on $[c, d]$. If $h(t)$ is close to λ_j for $t_{j-1} < t < t_j$, and the sums (1) approach a limit as the approximation to h gets closer, we call this limit the 'integral of h with respect to φ' (to be denoted by $\int h d\varphi$). Having given a precise meaning to this integral, we shall simply define $\int_\varphi f$ to be $\int (f \circ \varphi) d\varphi$.

In order to define $\int h d\varphi$, we will proceed by first making an obvious definition for step functions, and then extending the definition (in a continuous way) to functions that can be uniformly approximated by step functions. (For a slightly different method, see Duncan (2)).

When φ is piecewise-smooth, we find (1.7.4) that $\int h d\varphi$ is the same as the previously defined integral $\int_c^d h\varphi'$ (this is simply the integral of a complex function of a real variable, of the kind considered in 1.4). This is how such integrals are usually evaluated in practice. It is logically adequate to take $\int_c^d h\varphi'$ (whenever it exists) as the *definition* of $\int h d\varphi$ (thereby only defining integrals on piecewise-smooth paths). This approach is sufficient for the many applications of integration that appear in this book, but it can hardly be described as well-motivated. However, readers who wish to adopt this course may proceed straight to 1.7.5, after convincing themselves that 1.7.2 and suitable versions of properties (2)–(5) below still hold. (This is a very elementary exercise; paths 'equivalent' to φ should be restricted to mean paths of the form $\varphi \circ g$, where g is strictly increasing and *smooth*.)

Let $[c, d]$ be a real interval. A function h from $[c, d]$ to $\mathbf{C}$ is said to be a *step function* on $[c, d]$ if there exist t_j in $[c, d]$ and λ_j in $\mathbf{C}$ such that $c = t_0 < t_1 < \cdots < t_n = d$ and $h(t) = \lambda_j$ for $t_{j-1} < t < t_j$

(nothing is said about the values of h at the t_j). For such h, we define the (Stieltjes) integral of h with respect to φ to be

$$\int h\, d\varphi = \sum_{j=1}^{n} \lambda_j[\varphi(t_j) - \varphi(t_{j-1})].$$

Verification of the following properties is immediate:

(1) $|\int h\, d\varphi| \leqslant L(\varphi) \max |\lambda_j|$.

(2) If h and k are step functions on $[c,d]$, then so are αh ($\alpha \in \mathbf{C}$) and $h+k$, and $\int (\alpha h)\, d\varphi = \alpha \int h\, d\varphi$, $\int (h+k)\, d\varphi = \int h\, d\varphi + \int k\, d\varphi$.

(3) If $\varphi \circ g$ is a path equivalent to φ (sect. 1.4), then $\int (h \circ g)\, d(\varphi \circ g) = \int h\, d\varphi$.

(4) If $(-\varphi)(t) = \varphi(-t)$ $(-d \leqslant t \leqslant -c)$ and $h_1 t() = h(-t)$, then $\int h_1\, d(-\varphi) = -\int h\, d\varphi$.

(5) If $c < e < d$, then $\int h\, d\varphi = \int h\, d\varphi_1 + \int h\, d\varphi_2$, where φ_1, φ_2 are the restrictions of φ to $[c,e]$ and $[e,d]$.

The extension of our definition depends on the following lemma.

1.7.1 Lemma. *Let (X,d) be a metric space, and let S be a dense subset. Let f be a function from S to $\mathbf{C}$ such that, for some $k>0$, $|f(s)-f(t)| \leqslant k d(s,t)$ for all s, t in S. Then there is a unique continuous function $\bar{f}$ from X to $\mathbf{C}$ that extends f, and $|\bar{f}(x) - \bar{f}(x')| \leqslant k d(x,x')$ for all x, x' in X.*

Proof. Take x, x' in X, and write $d(x,x') = r$. There exist sequences $\{s_n\}$, $\{s'_n\}$ in S that converge to x, x' respectively. Then the sequences $\{f(s_n)\}$, $\{f(s'_n)\}$ are Cauchy, so have limits y, y' in $\mathbf{C}$. We prove that $|y-y'| \leqslant kr$. Take $\epsilon > 0$. There exists N such that for all $n \geqslant N$, $d(s_n, x) \leqslant \epsilon/2$ and $d(s'_n, x') \leqslant \epsilon/2$. Therefore for m, $n \geqslant N$, $d(s_m, s'_n) \leqslant r + \epsilon$, so $|f(s_m) - f(s'_n)| \leqslant k(r+\epsilon)$. Letting $n \to \infty$, this shows that $|f(s_m) - y'| \leqslant k(r+\epsilon)$ $(m \geqslant N)$, and now letting $m \to \infty$, we have $|y-y'| \leqslant k(r+\epsilon)$. This is true for all $\epsilon > 0$, so $|y-y'| \leqslant kr$ (1).

If $x = x'$, then $r = 0$, so (1) shows that $y = y'$. Hence we can define a function $\bar{f}$ on X by putting $\bar{f}(x) = \lim_{n\to\infty} f(s_n)$, where $\{s_n\}$ is any sequence in S that converges to x. For this function, (1) now shows that $|\bar{f}(x) - \bar{f}(x')| \leqslant k d(x,x')$ for all x, x' in X. Clearly, $\bar{f}$ is the only continuous function on X that extends f.

Let $S[c,d]$ denote the set of all complex-valued step functions on $[c,d]$, and $B[c,d]$ the set of all bounded complex-valued functions on $[c,d]$. We give $B[c,d]$ the usual supremum norm:

$$\|h\| = \sup\{|h(t)| : c \leqslant t \leqslant d\}.$$

Let $I[c,d]$ be the closure (with respect to this norm) of $S[c,d]$ in $B[c,d]$. Then h is in $I[c,d]$ if and only if, for each $\epsilon > 0$, there is a member k_ϵ of $S[c,d]$ such that $\|h - k_\epsilon\| \leqslant \epsilon$. In other words, $I[c,d]$ is the set of functions on $[c,d]$ that can be uniformly approximated by step functions.

If h and k are in $S[c,d]$, then properties (1) and (2) above show that

$$\left|\int h\,d\varphi - \int k\,d\varphi\right| \leqslant L(\varphi)\|h - k\|.$$

We can therefore use 1.7.1 to extend the mapping $h \mapsto \int h\,d\varphi$, increasing its domain from $S[c,d]$ to $I[c,d]$. The value assigned to h by the extended mapping is again denoted by $\int h\,d\varphi$, and referred to as the integral of h with respect to φ.

Finally, if f is a complex-valued function with domain including φ^*, and $f \circ \varphi \in I[c,d]$, we say that f is *integrable on* φ, and define $\int_\varphi f$ to be $\int (f \circ \varphi)\,d\varphi$. When it is more convenient, we write $\int_\varphi f(z)\,dz$ instead of $\int_\varphi f$ (see the introductory section on Terminology and Notation).

1.7.1 also gives us the following important inequality:

1.7.2. *For h, k in $I[c,d]$,* $\left|\int h\,d\varphi - \int k\,d\varphi\right| \leqslant L(\varphi)\|h - k\|$.
If f and g are integrable on φ, then

$$\left|\int_\varphi f - \int_\varphi g\right| \leqslant L(\varphi) \sup\{|f(z) - g(z)| : z \in \varphi^*\}.$$

Hence if $f_n \to f$ uniformly on φ^*, each f_n being integrable on φ, then $\int_\varphi f = \lim_{n\to\infty} \int_\varphi f_n$. ($f$ is integrable on φ, since $f_n \circ \varphi \to f \circ \varphi$ uniformly on $[c,d]$, and $I[c,d]$ is closed, by construction.)

Another useful consequence of 1.7.2 is the following: if we have approximated to h by a step function k, and $\|h - k\| \leqslant \epsilon$, then $\left|\int h\,d\varphi - \int k\,d\varphi\right| \leqslant \epsilon L(\varphi)$.

It is a trivial matter to deduce from (2)–(5) above that similar statements are true for functions in $I[c,d]$. The corresponding statements for complex functions are:

(2′) $\int_\varphi \alpha f = \alpha \int_\varphi f$, $\int_\varphi (f+g) = \int_\varphi f + \int_\varphi g$.
(3′) If ψ is equivalent to φ, then $\int_\psi f = \int_\varphi f$.
(4′) $\int_{-\varphi} f = -\int_\varphi f$.
(5′) If $\varphi \cup \psi$ is defined, then $\int_{\varphi\cup\psi} f = \int_\varphi f + \int_\psi f$.

1.7.3. *If h is continuous on $[c,d]$, then $h \in I[c,d]$. If f is continuous on φ^*, then f is integrable on φ.*

Proof. The second statement follows from the first, since $f \circ \varphi$ is continuous. Take $\epsilon > 0$. Since h is uniformly continuous, there exists $\delta > 0$ such that if $t, u \in [c,d]$ and $|t-u| \leqslant \delta$, then $|h(t) - h(u)| \leqslant \epsilon$. Take a dissection $c = t_0 < t_1 < \cdots < t_n = d$ with $t_j - t_{j-1} \leqslant \delta$ for each j, and let k be the step function taking the constant value $h(t_j)$ on $[t_{j-1}, t_j)$ for each j. Then $\|h-k\| \leqslant \epsilon$.

Examples. (i) Let $f(z) = 1$ for all z, and let φ be any rectifiable path from a to b. Then $\int_\varphi f = b - a$.

(ii) If $f(z) = z$, then $f \circ \varphi = \varphi$. Hence $\int_\varphi z\,dz = \int \varphi\,d\varphi$. We show that $\int \varphi\,d\varphi = \frac{1}{2}(b^2 - a^2)$ for any rectifiable path φ from a to b. Take $\epsilon > 0$. Since φ is uniformly continuous, there is a dissection $c = t_0 < t_1 < \cdots < t_n = d$ such that $|\varphi(t) - \varphi(u)| \leqslant \epsilon$ whenever $t, u \in [t_{j-1}, t_j]$. Write $z_j = \varphi(t_j)$, and let

$$s_1 = a(z_1 - a) + z_1(z_2 - z_1) + \cdots + z_{n-1}(b - z_{n-1}),$$

$$s_2 = z_1(z_1 - a) + z_2(z_2 - z_1) + \cdots + b(b - z_{n-1}).$$

Then $s_j = \int_\varphi k_j$ $(j = 1,2)$, where k_1, k_2 are step functions such that $\|\varphi - k_j\| \leqslant \epsilon$. Therefore $|\int \varphi\,d\varphi - s_j| \leqslant \epsilon L(\varphi)$ for $j = 1$, 2. Since $s_1 + s_2 = b^2 - a^2$, the result follows.

Notice that, in both these examples, $\int_\varphi f$ depends only on the end-points of φ.

Evaluation of integrals on a piecewise-smooth path

1.7.4 Theorem. *If φ is a piecewise-smooth path in* $\mathbf{C}$, *with domain $[c,d]$, and $h \in I[c,d]$, then*

$$\int h\,d\varphi = \int_c^d h\varphi'.$$

Proof. By property (5), it s sufficient to prove the result in the case when φ is smooth. The functions h and φ' are bounded on $[c,d]$: let $M = \max[\|h\|, \|\varphi'\|, L(\varphi)]$. Take $\epsilon > 0$. Now h can be approximated by step functions, and φ' is uniformly continuous on $[c,d]$. It follows that we can choose a dissection $c = t_0 < t_1 < \cdots < t_n = d$ and numbers $\lambda_1, \ldots, \lambda_n$ such that, for each j,

(i) $|\lambda_j| \leqslant M$,

(ii) $|\varphi'(t) - \varphi'(u)| \leqslant \epsilon/M$ for t, u in $[t_{j-1}, t_j]$,

(iii) $|h(t) - \lambda_j| \leqslant \epsilon/M$ for t in $[t_{j-1}, t_j]$.

The third condition says that if k is the step function that takes the value λ_j on $[t_{j-1}, t_j)$, then $|h(t) - k(t)| \leqslant \epsilon/M$ $(c \leqslant t \leqslant d)$. Hence, writing $\varphi(t_j) = z_j$, we have

$$\left|\int h\,d\varphi - \sum_{j=1}^{n} \lambda_j(z_j - z_{j-1})\right| \leqslant \epsilon.$$

Let $\varphi(t) = p(t) + iq(t)$, where $p(t)$ and $q(t)$ are real, and write

$$a_j = \frac{z_j - z_{j-1}}{t_j - t_{j-1}}.$$

By the mean value theorem, there exist u_j, v_j in (t_{j-1}, t_j) such that $a_j = p'(u_j) + iq'(v_j)$. For $t_{j-1} \leqslant t \leqslant t_j$, (ii) shows that $|p'(t) - p'(u_j)| \leqslant \epsilon/M$, and similarly for q, so $|\varphi'(t) - a_j| \leqslant 2\epsilon/M$, and

$$|h(t)\,\varphi'(t) - a_j\lambda_j| \leqslant |(h(t) - \lambda_j)\,\varphi'(t)| + |\lambda_j(\varphi'(t) - a_j)|$$

$$\leqslant \frac{\epsilon}{M}M + M\frac{2\epsilon}{M} = 3\epsilon.$$

In other words, if ψ is the step function that takes the value $a_j\lambda_j$ on $[t_{j-1}, t_j)$, then $|(h\varphi')(t) - \psi(t)| \leqslant 3\epsilon$ $(c \leqslant t \leqslant d)$. Hence $|\int_c^d h\varphi' - \int_c^d \psi| \leqslant 3\epsilon(d-c)$. The result follows, since

$$\int_c^d \psi = \sum_{j=1}^{n} a_j\lambda_j(t_j - t_{j-1}) = \sum_{j=1}^{n} \lambda_j(z_j - z_{j-1}).$$

An important consequence of this theorem is the fact that the integral of a derivative behaves in the same way as for real functions:

1.7.5 Corollary. *If φ is a piecewise-smooth path from a to b, and f is a smooth complex function defined on an open set containing φ^*, then*

$$\int_\varphi f' = f(b) - f(a).$$

Proof. By the theorem, $\int_\varphi f' = \int_c^d (f' \circ \varphi)\varphi'$. By 1.5.4, $(f' \circ \varphi)\varphi' = (f \circ \varphi)'$. Therefore (using 1.4.3),

$$\begin{aligned}\int_\varphi f' &= (f \circ \varphi)(d) - (f \circ \varphi)(c)\\ &= f(b) - f(a).\end{aligned}$$

1.7.5 is, in fact, true for all rectifiable paths (and the same therefore applies to the later results that depend on it). The proof is a routine application of compactness and the definition of differentiability, and the reader may care to attempt it as an exercise.

1.7.5 has an implication for complex functions that is entirely lacking in the case of real functions. Suppose that φ and ψ are two

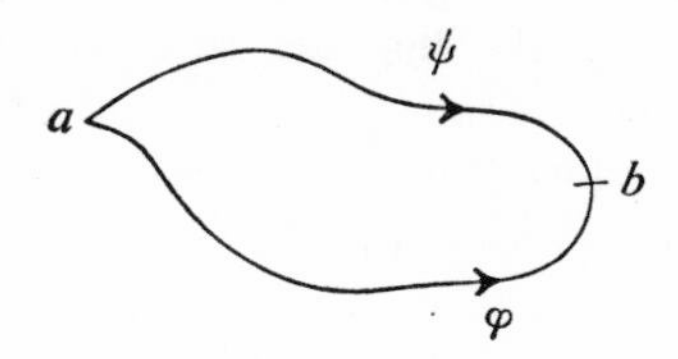

piecewise-smooth paths from a to b. Then $\varphi \cup (-\psi)$ is a closed, piecewise-smooth path, and for any integrable function g,

$$\int_{\varphi \cup (-\psi)} g = \int_\varphi g - \int_\psi g.$$

Hence if G is a connected open set, and g is a continuous function from G to $\mathbf{C}$, then the following two statements are equivalent:

(i) If $a, b \in G$ and φ, ψ are two piecewise-smooth paths in G from a to b, then $\int_\varphi g = \int_\psi g$.

(ii) $\int_\varphi g = 0$ for every closed, piecewise-smooth path φ in G.

1.7.5 shows that (i) and (ii) hold for any continuous function g that is the derivative of another function on G. In chapter 2 we will prove that if G is star-shaped, then (i) and (ii) hold for every differentiable function on G.

Examples. (i) To integrate e^z along the directed line segment $[0 \to i\pi/2]$. Since $e^z = (d/dz)e^z$, 1.7.5 shows that the integral is

$\exp(i\pi/2) - e^0 = i - 1$. The same answer is obtained if we integrate along any rectifiable path from 0 to $i\pi/2$.

(ii) Let $f(z) = |z|^2$. We find the integral of f along two different paths from 1 to i: (a) $[1 \to i]$, (b) the arc of the unit circle, i.e. $t \mapsto e^{it}$ $(0 \leqslant t \leqslant \pi/2)$.

(a) The path is φ, where $\varphi(t) = (1-t) + it$ $(0 \leqslant t \leqslant 1)$. Now $\varphi'(t) = i - 1$ and $f(\varphi(t)) = (1-t)^2 + t^2$, so the integral is

$$(i-1)\int_0^1 ((1-t)^2 + t^2)\,dt = \tfrac{2}{3}(i-1).$$

(b) The function is equal to 1 on the path, so the integral is $i - 1$.

(iii) Often the computation is simplified if we use an equivalent path in evaluating an integral (we may do so, by elementary property (3) above). To illustrate this, we integrate f along $[1 - i \to 1 + i]$, where $f(x + iy) = xy^2$. Let $\varphi(t) = 1 + it$ $(-1 \leqslant t \leqslant 1)$. Then φ is an equivalent path, and $\varphi'(t) = i$. Since $f(1 + it) = t^2$, the integral is $i\int_{-1}^1 t^2\,dt = \tfrac{2}{3}i$.

Recall that the $C(a,r)$ denotes the path $t \mapsto a + re^{it}$ $(0 \leqslant t \leqslant 2\pi)$. The following special case is important enough to be made into a theorem:

1.7.6. *Take $r > 0$. Then the integral $\int_{C(a,r)} (z-a)^n\,dz$ is 0 if n is an integer different from -1, and*

$$\int_{C(a,r)} \frac{1}{z-a}\,dz = 2\pi i.$$

Proof. If $n \neq -1$, then $(z-a)^n$ is a derivative away from a, so the first statement follows from 1.7.5. Also,

$$\int_{C(a,r)} \frac{1}{z-a}\,dz = \int_0^{2\pi} \frac{1}{re^{it}}\,ir\,e^{it}\,dt = 2\pi i.$$

Integral along a transformed path

If φ is a piecewise-smooth path in $\mathbf{C}$, and f is a smooth complex-valued function defined on an open set containing φ^*, then $f \circ \varphi$ is a piecewise-smooth path, since $(f \circ \varphi)' = (f' \circ \varphi)\varphi'$. The next result shows how an integral along $f \circ \varphi$ can be expressed as an integral along φ.

1.7.7. *If f, φ are as above, and g is integrable on $f \circ \varphi$, then $\int_{f \circ \varphi} g = \int_\varphi (g \circ f) f'$.*

Proof. $\int_{f \circ \varphi} g = \int_c^d g \circ (f \circ \varphi)(f \circ \varphi)'$, by 1.7.4,

$$= \int_c^d (g \circ f) \circ \varphi . (f' \circ \varphi) \varphi'$$

$$= \int_c^d (h \circ \varphi) \varphi'$$

$$= \int_\varphi h,$$

where $h = (g \circ f) f'$.

Example. Putting $g(z) = \bar{z}$, this gives

$$\int_\varphi \bar{f} f' = \int_{f \circ \varphi} \bar{z}\, dz.$$

Integration of power series

We have seen in section 1.5 that the coefficients in a power series can be expressed as derivatives of the sum-function, a result that applies equally to real and complex series. We now show that, in the complex case, the coefficients can also be expressed as integrals involving the sum-function. This has no counterpart in the real case.

1.7.8. *Suppose that $\sum_{n=0}^\infty a_n(z - z_0)^n = f(z)$ for $|z - z_0| < R$, where $R > 0$. Then*

$$a_n = \frac{1}{2\pi i} \int_{C(z_0, r)} \frac{f(z)}{(z - z_0)^{n+1}} dz \qquad (n = 0, 1, 2, \ldots),$$

for any r in $(0, R)$.

Proof. For fixed n,

$$\frac{f(z)}{(z - z_0)^{n+1}} = \sum_{p=0}^\infty a_p (z - z_0)^{p-n-1},$$

convergence being uniform on $C(z_0,r)^*$. By 1.7.2, it follows that

$$\int_{C(z_0,r)} \frac{f(z)}{(z-z_0)^{n+1}}\,dz = \sum_{p=0}^{\infty} a_p \int_{C(z_0,r)} (z-z_0)^{p-n-1}\,dz.$$

But the right-hand side is equal to $2\pi i a_n$, by 1.7.6.

In particular, we notice that

$$f(z_0) = a_0 = \frac{1}{2\pi i}\int_{C(z_0,r)} \frac{f(z)}{z-z_0}\,dz.$$

1.7.9 Corollary. *Suppose that* $\sum_{n=0}^{\infty} a_n(z-z_0)^n = f(z)$ *for* $|z-z_0| < R$, *where* $R > 0$. *For* $0 < r < R$, *let* $M(r) = \sup\{|f(z)| : |z-z_0| = r\}$. *Then* $M(r) \geqslant |a_n|\, r^n$ *for each* n.

Proof. Since the length of $C(z_0,r)$ is $2\pi r$, the expression in 1.7.8 gives

$$|a_n| \leqslant \frac{1}{2\pi}\frac{M(r)}{r^{n+1}}2\pi r = \frac{M(r)}{r^n}.$$

Clearly, we also have the converse inequality

$$M(r) \leqslant \sum_{n=0}^{\infty} |a_n|\, r^n.$$

The reader may have noticed that integration is not mentioned in the statement of 1.7.9. In fact, this is a result which we have proved with the aid of integration, but which could be stated even if integration had never been invented. We shall have many further examples of such results in chapter 2.

Exercises 1.7

1. Find the integrals of $z(z-1)$ and $\operatorname{Re} z$ along the line segments $[0 \to 1+i]$, $[0 \to 1]$ and $[1 \to 1+i]$.

2. Let $f(x+iy) = xy$. Show that the integral of f along the semicircle $t \mapsto e^{it}$ $(0 \leqslant t \leqslant \pi)$ is $\frac{2}{3}i$.

3. Find the integral of $1/z$ round the square with vertices $\pm 1 \pm i$ in the anticlockwise direction, i.e. along the succession

of line segments $[1-i \to 1+i]$, $[1+i \to -1+i]$, $[-1+i \to -1-i]$, $[-1-i \to 1-i]$. (Hint: combine the integrals for opposite sides before evaluating them.)

4 Suppose that $f(z) = \sum_{n=0}^{\infty} a_n z^n$ for $|z| < R$, where $R > 0$. Prove that if $0 < r < R$ and $n > 0$, then there exists z such that $|z| = r$ and

$$\left|\int_{[0\to z]} f\right| \geqslant \frac{1}{n}|a_{n-1}|\, r^n.$$

5 By taking points of dissection $2\pi ri/n$ $(r = 0, 1, \ldots, n-1)$, prove directly that the integral of $1/z$ round the unit circle is $2\pi i$. (Use exercise 3 of 1.6.)

6 If φ is a closed path, show that $\int_\varphi \bar{z}\, dz$ is purely imaginary:
(i) for rectifiable paths, by adding two approximating sums,
(ii) for piecewise-smooth paths, by using 1.7.4.

7 If f is an even function, show that $\int_{C(0,\, r)} f = 0$ for all $r > 0$.

8 Show by integration that the series expression given in 1.6.17 for $\log_0(1+z)$ is also valid for z such that $|z| = 1$ and $z \neq -1$. Deduce that

$$\sum_{n=1}^{\infty} \frac{(-1)^{n+1}}{n} \sin nt = \tfrac{1}{2}t$$

for $-\pi < t < \pi$.

CHAPTER TWO

The theory of differentiable functions

In this chapter we investigate the properties of complex functions that are assumed to be differentiable, not just at an isolated point, but throughout an open set. We obtain a spectacular sequence of results, most of them quite unlike any theorems applying to real functions. This is one of the classic theories of Mathematics – there is perhaps no other instance in which a comparable succession of good theorems is obtained with such ease and elegance.

2.1. Cauchy's integral theorem and formula

Cauchy's integral theorem

Cauchy's integral theorem is the result, already mentioned in 1.7, that if a complex function is differentiable on a suitably shaped open set G, then $\int_{\varphi} f = 0$ for every closed, piecewise-smooth path φ in G. We have seen (1.7.5) that this is true if f is the derivative of another function, and the scheme of our proof is to show that every differentiable function on G is, in fact, the derivative of another function. For real functions, this is very easy to prove (and only continuity need be assumed). One defines $F(x) = \int_a^x f$ (for some fixed a), and shows that $F' = f$. The proof is elementary on noticing that $F(x) = F(x_0) + \int_{x_0}^x f$. We start by showing that a similar result holds for complex functions if we replace the interval $[x_0, x]$ by a line segment $[z_0 \to z]$.

2.1.1. *Suppose that f is continuous and that $F(z) = F(z_0) + \int_{[z_0 \to z]} f$ or z in $D(z_0, r)$ (where $r > 0$). Then $F'(z_0) = f(z_0)$.*

Proof. Take $\epsilon > 0$. There exists δ in $(0, r)$ such that $|f(\zeta) - f(z_0)| \leqslant \epsilon$ whenever $|\zeta - z_0| \leqslant \delta$. Write $g(\zeta) = f(\zeta) - f(z_0)$. Then

$$F(z) - F(z_0) = \int_{[z_0 \to z]} f$$
$$= (z - z_0) f(z_0) + \int_{[z_0 \to z]} g.$$

If $|z - z_0| \leqslant \delta$, then $|\int_{[z_0 \to z]} g| \leqslant \epsilon |z - z_0|$, and the result follows.

Returning to the problem of showing that a given function f is the derivative of another function on G, let a be a point of G, and for z in G, let $F(z)$ be the integral of f along some path from a to z. It is natural to choose the simplest possible path, viz. the line segment $[a \to z]$. To ensure that this line segment lies in G, we must assume that G is star-shaped about a. Then 2.1.1 shows that $F'(a) = f(a)$, but to conclude that $F'(z_0) = f(z_0)$ for another point z_0 of G, we need to know that for z sufficiently close to z_0, $F(z) - F(z_0) = \int_{[z_0 \to z]} f$. This condition can be rewritten

$$\int_{[a \to z_0]} f + \int_{[z_0 \to z]} f + \int_{[z \to a]} f = 0.$$

It is now clear that the condition we want is that (loosely speaking) the integral of f round every triangle is zero. To make the precise statement manageable, we introduce some terminology concerning triangles.

By a **triangle** in $\mathbf{C}$ we mean an ordered triple $T = (a, b, c)$ of points of $\mathbf{C}$. We denote by $\operatorname{co}(T)$ the convex cover of $\{a, b, c\}$, and recall that by 1.3.3, $\operatorname{co}(T)$ is compact. We denote by ∂T the closed polygonal path $[a \to b] \cup [b \to c] \cup [c \to a]$. The result suggested by the analysis above is:

2.1.2. *Let G be an open subset of $\mathbf{C}$, star-shaped about a. Let f be a continuous function from G to $\mathbf{C}$, and suppose that for every triangle T with $\operatorname{co}(T) \subseteq G$, we have $\int_{\partial T} f = 0$. Define $F(z) = \int_{[a \to z]} f \quad (z \in G)$. Then $F'(z) = f(z) \quad (z \in G)$.*

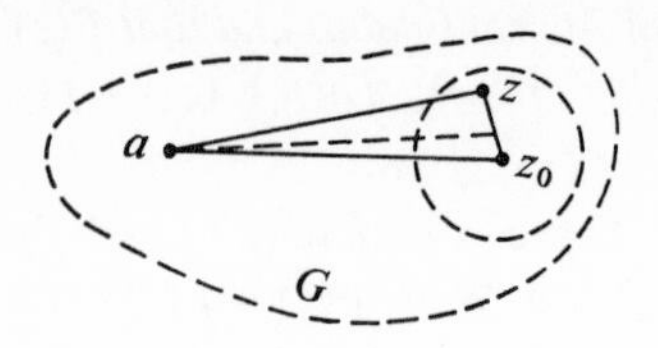

Proof. Take z_0 in G. There exists $r > 0$ such that $D(z_0, r) \subseteq G$. Suppose that $z \in D(z_0, R)$. Then G contains the line segment $[z_0 : z]$. Now each point of $\text{co}\{a, z_0, z\}$ lies on a line segment from a to a point of $[z_0 : z]$, so $\text{co}\{a, z_0, z\} \subseteq G$. Therefore, by the hypothesis on integration round triangles, $F(z) = F(z_0) + \int_{[z_0 \to z]} f$, and 2.1.1 shows that $F'(z_0) = f(z_0)$.

We are now in a position to prove Cauchy's integral theorem.

2.1.3 Theorem. *Let G be a star-shaped open subset of* $\mathbf{C}$, *and let f be a differentiable function from G to* $\mathbf{C}$. *Then* (i) *there is a function F on G such that $F' = f$, and* (ii) *for every closed, piecewise-smooth path φ in G, $\int_\varphi f = 0$.*

Proof. It is sufficient to prove that f satisfies the condition of 2.1.2. Let $T = (a, b, c)$ be a triangle such that $\text{co}(T) \subseteq G$. We must show that $\int_{\partial T} f = 0$. Let L be the length of ∂T, that is, $|b - a| + |c - b| + |a - c|$, and for every triangle T' constructed, write $\eta(T') = \int_{\partial T'} f$.

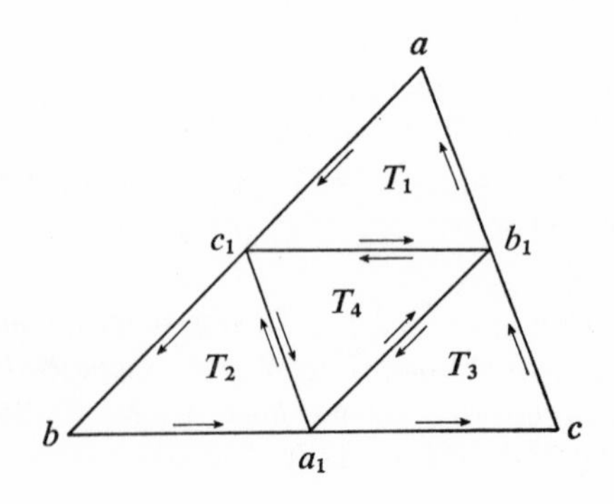

We quadrisect T, in the following carefully defined sense. Write $a_1 = \frac{1}{2}(b+c)$, $b_1 = \frac{1}{2}(c+a)$, $c_1 = \frac{1}{2}(a+b)$. Let $T^{(j)}$ $(j = 1,2,3,4)$ be the triangles (a,c_1,b_1), (b,a_1,c_1), (c,b_1,a_1), (a_1,b_1,c_1) respectively. Then the length of each $\partial T^{(j)}$ is $\frac{1}{2}L$, and, by cancellations, $\eta(T) = \sum_{j=1}^{4} \eta(T^{(j)})$. Therefore for at least one of the $T^{(j)}$, to be denoted by T_1, we have $|\eta(T_1)| \geqslant \frac{1}{4}|\eta(T)|$.

Now quadrisect T_1 in the same way. For one of the four triangles obtained, to be denoted by T_2, we have $|\eta(T_2)| \geqslant \frac{1}{4}|\eta(T_1)|$. Repeating the process, we obtain a sequence $\{T_n\}$ of triangles such that, for each n: (i) $|\eta(T_n)| \geqslant 4^{-n}|\eta(T)|$, (ii) the length of ∂T_n is $L/2^n$, (iii) $\operatorname{co}(T_n) \subseteq \operatorname{co}(T_{n-1})$.

Now $\operatorname{co}(T_n)$ is compact for each n, so there is a point z_0 in

$$\bigcap_{n=1}^{\infty} \operatorname{co}(T_n)$$

(by 0.4). Take $\epsilon > 0$. Since f is differentiable at z_0, there exists $\delta > 0$ such that for $|z - z_0| \leqslant \delta$,

$$|f(z) - f(z_0) - (z - z_0) f'(z_0)| \leqslant \epsilon |z - z_0|.$$

Take n such that $L/2^n < \delta$. By 1.3.2, the diameter of $\operatorname{co}(T_n)$ is certainly not greater than $L(\partial T_n)$, so $|z - z_0| < \delta$ for z in $\operatorname{co}(T_n)$. Therefore, by 1.7.2,

$$\left|\int_{\partial T_n} (f(z) - f(z_0) - (z - z_0) f'(z_0))\, dz\right| \leqslant \epsilon \frac{L}{2^n} \frac{L}{2^n}.$$

But

$$\int_{\partial T_n} (f(z_0) + (z - z_0) f'(z_0))\, dz = 0,$$

since the integrand is a derivative. Hence $|\eta(T_n)| \leqslant \epsilon L^2/4^n$, so $|\eta(T)| \leqslant \epsilon L^2$. This is true for all $\epsilon > 0$, so $\eta(T) = 0$.

With a lot more work, it is possible to prove Cauchy's integral theorem for rather more general sets (see Mackey [3], page 117, or Ahlfors [1], page 141), but the form given here is sufficient for the purposes of this book.

Integration round points where f is not differentiable

The next theorem – our first deduction from Cauchy's theorem – is fundamental for much of our further theory. Loosely speaking, it says that if f is differentiable except on a bounded set, and two circles (one inside the other) both enclose this set, then the integral of f round each circle is the same. The proof may appear slightly involved, but the basic idea is very simple, and the complication only arises in being precise about constructions which seem obvious on a diagram.

2.1.4 Theorem. *Let z_1, z_2 be complex numbers and ρ_1, ρ_2 positive numbers such that $\rho_1 > \rho_2 + |z_1 - z_2|$. Suppose that there exist $r_1 > \rho_1$ and $r_2 < \frac{1}{2}\rho_2$ such that f is differentiable on*

$$\{z : |z - z_1| < r_1\} \cap \{z : |z - z_2| > r_2\}.$$

Write $C(z_j, \rho_j) = C_j$ $(j = 1,2)$. Then

$$\int_{C_1} f = \int_{C_2} f.$$

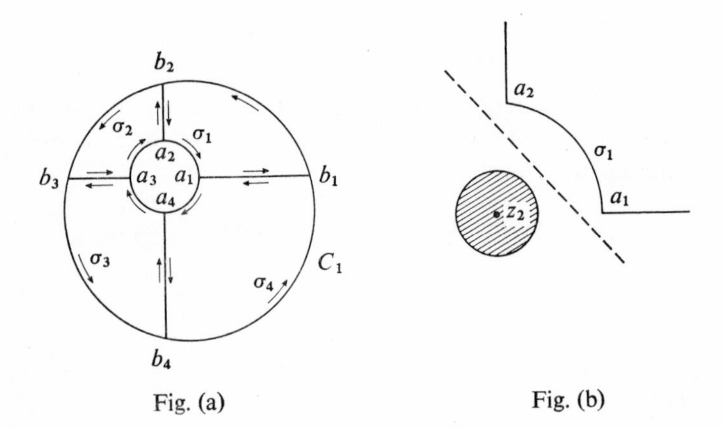

Fig. (a) Fig. (b)

Proof. Let $z_j = x_j + iy_j$ $(j = 1,2)$, and let a_1, b_1 be the points on C_2^*, C_1^* respectively of the form $x' + iy_2$, where $x' > x_2$. (Clearly, $a_1 = x_2 + \rho_2 + iy_2$; the real part of b_1 is found by solving $(x' - x_1)^2 + (y_2 - y_1)^2 = \rho_1^2$, but its existence is all that interests us.) Define a_k, b_k $(k = 2,3,4)$ similarly, as indicated in Fig. (a). Let σ_1 denote the closed path consisting of $[a_1 \to b_1]$, the part of C_1 between b_1 and b_2 (i.e. with $x \geqslant x_2$ and $y \geqslant y_2$), $[b_2 \to a_2]$, and the part of $-\varphi_2$ between a_2 and a_1 (i.e. $t \mapsto z_2 + \rho_2 e^{-it}$ for $-\pi/2 \leqslant t \leqslant 0$). Define σ_2, σ_3, σ_4 similarly, as indicated in Fig. (a).

Let G_1 be the set of $z = x + iy$ such that $|z - z_1| < r_1$ and $x + y > x_2 + y_2 + 2r_2$. Then G_1 is open and convex. We show that σ_1 lies in G_1 and that f is differentiable on G_1, from which it will follow that $\int_{\sigma_1} f = 0$. For $z = x + iy$ in σ_1^*, we have $x \geqslant x_2$, $y \geqslant y_2$ and $|z - z_2| > \rho_2 > 2r_2$, so that $(x - x_2) + (y - y_2) \geqslant |z - z_2| > 2r_2$, and $z \in G_1$. On the other hand, if $z = x + iy$ and $|z - z_2| \leqslant r_2$, then $x - x_2 \leqslant r_2$ and $y - y_2 \leqslant r_2$, so $z \notin G_1$. Hence f is differentiable on G_1.

Similarly, $\int_{\sigma_k} f = 0$ for $k = 2, 3, 4$. But, by cancellations along the line segments,

$$\sum_{k=1}^{4} \int_{\sigma_k} f = \int_{C_1} f - \int_{C_2} f.$$

As a first application of this theorem, we note:

2.1.5 Corollary. *The integral*

$$\int_{C(a,\, r)} \frac{1}{z - z_0}\, dz$$

has the value $2\pi i$ *if* $|z_0 - a| < r$, *and* 0 *if* $|z_0 - a| > r$.

Proof. Write $f(z) = 1/(z - z_0)$ $(z \neq z_0)$. If $|z_0 - a| > r$, choose ρ such that $|z_0 - a| > \rho > r$. Then f is differentiable on the convex, open set $D(a, \rho)$, which contains $C(a, r)$, so the integral is zero, by 2.1.3. If $|z_0 - a| < r$, choose ρ such that $0 < \rho < r - |z_0 - a|$. By 2.1.4, $\int_{C(a,\, r)} f = \int_{C(z_0,\, \rho)} f$. But, by 1.7.6, $\int_{C(z_0,\, \rho)} f = 2\pi i$.

It would make no difference to the proof of 2.1.4 if the outer circle were replaced by various other closed paths, for instance a rectangle (cf. exercise 3 of 1.7). This remark becomes important later (section 3.1). However, we will not attempt to describe a general class of paths to which it applies, preferring to recognize its truth in particular cases as they arise.

The reader may have suspected that the condition $r_2 < \frac{1}{2}\rho_2$ in 2.1.4 is unnecessarily restrictive. This is, indeed, the case; for the details, see exercise 5.

Cauchy's integral formula

2.1.6 Theorem. *Suppose that f is differentiable on $D(z_0, R)$, and that $0 < r < R$. Then, for all z in $D(z_0, r)$,*

$$f(z) = \frac{1}{2\pi i} \int_{C(z_0,\, r)} \frac{f(\zeta)}{\zeta - z} d\zeta.$$

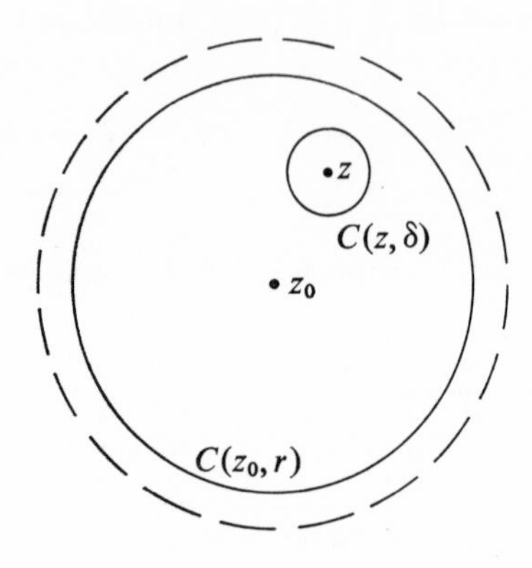

Proof. Choose z in $D(z_0, r)$. For ζ in $D(z_0, R) \backslash \{z\}$, define

$$F(\zeta) = \frac{f(\zeta)}{\zeta - z}.$$

Then F is differentiable where defined. Take $\epsilon > 0$. There exists δ such that $0 < \delta < r - |z - z_0|$ and that $|f(\zeta) - f(z)| \leqslant \epsilon$ whenever $|\zeta - z| \leqslant \delta$. By 2.1.4, $\int_{C(z_0,\, r)} F = \int_{C(z,\, \delta)} F$. But

$$\int_{C(z,\, \delta)} F = f(z) \int_{C(z,\, \delta)} \frac{1}{\zeta - z} d\zeta + \int_{C(z,\, \delta)} g,$$

where

$$g(\zeta) = \frac{f(\zeta) - f(z)}{\zeta - z}.$$

Now $|g(\zeta)| \leqslant \epsilon/\delta$ for ζ in $C(z, \delta)^*$, so

$$\left| \int_{C(z,\, \delta)} F - 2\pi i f(z) \right| = \left| \int_{C(z,\, \delta)} g \right| \leqslant \frac{\epsilon}{\delta} 2\pi\delta = 2\pi\epsilon,$$

by 1.7.2. This is true for all $\epsilon > 0$, so $\int_{C(z_0,\, r)} F = 2\pi i f(z)$.

2.1.7 Corollary. *If f is differentiable on $D(z_0, R)$, and $0 < r < R$, then*

$$f(z_0) = \frac{1}{2\pi} \int_0^{2\pi} f(z_0 + r e^{it})\, dt.$$

2.1.6 shows that, once the values of a differentiable complex function are known on the circumference of a circle, they are known at all points inside. The next three sections of this chapter consist of deductions from this remarkable result. There is, of course, no corresponding true statement for real functions: knowledge of the value of a differentiable real function at two points $x_0 - r$ and $x_0 + r$ certainly does not tell us anything about its values in the interval between them.

Cauchy's integral formula sometimes enables us to write down the value of an integral at sight. As an example, we evaluate

$$\int_{C(0,\,1)} \frac{1}{(z-a)^3 (z-b)}\, dz,$$

where $|a| > 1$ and $|b| < 1$. Let $f(z) = 1/(z-a)^3$. Then f is differentiable on $D(0, |a|)$, so

$$\int_{C(0,\,1)} \frac{f(z)}{z-b}\, dz = 2\pi i f(b) = \frac{2\pi i}{(b-a)^3}.$$

Exercises 2.1

1 Let $f(x+iy) = x+y$, and for z in $\mathbf{C}$, define $F(z) = \int_{[0\to z]} f$. At which points is F differentiable?

2 Suppose that φ is a closed, piecewise-smooth path in $\mathbf{C}$, and that f is a smooth complex-valued function defined on an open set containing φ^*. If $\{f(z) : z \in \varphi^*\}$ does not meet $\{x \in \mathbf{R} : x \leqslant 0\}$, prove that $\int_\varphi (f'/f) = 0$.

3 Evaluate

$$\int_{C(0,\,1)} \frac{1}{(z-a)(z-b)}\, dz,$$

where (i) $|a|, |b| < 1$, (ii) $|a| < 1, |b| > 1$, (iii) $|a|, |b| > 1$.

4 Evaluate $\int_{C(0,2)} \frac{e^z}{z-1}\,dz$ and $\int_{C(0,2)} \frac{e^z}{\pi i - 2z}\,dz$.

5 If $\omega_n = \exp(2\pi i/n)$, show that, for each k, the distance from 0 to the line through ω_n^{k-1} and ω_n^k is $\operatorname{Im}\omega_n/|1-\omega_n|$. Prove that this distance tends to 1 as $n \to \infty$. By constructing a suitably large number of closed paths (instead of four), show that the condition $r_2 < \rho_2$ is sufficient in 2.1.4 (instead of $r_2 < \frac{1}{2}\rho_2$).

2.2. The Taylor series and its applications

The Taylor series

Equipped with Cauchy's integral formula, we now derive the result (promised in 1.2) that a function that is differentiable on an open disc is the sum of a power series there. The scheme of the proof is very simple: we expand $1/(\zeta - z)$ as a geometric series and integrate term by term.

2.2.1 Theorem. *If f is differentiable on $D(z_0, R)$, then there exist unique complex numbers a_n $(n = 0, 1, 2, \ldots)$ such that*

$$f(z) = \sum_{n=0}^{\infty} a_n (z - z_0)^n$$

for z in $D(z_0, R)$. The a_n are given by

$$a_n = \frac{1}{2\pi i} \int_{C(z_0, r)} \frac{f(\zeta)}{(\zeta - z_0)^{n+1}}\,d\zeta \qquad (n \geqslant 0;\ 0 < r < R).$$

Proof. Choose z in $D(z_0, R)$, and write $|z - z_0| = \rho$. Take r such that $\rho < r < R$. By 2.1.6, we have

$$f(z) = \frac{1}{2\pi i} \int_{C(z_0, r)} \frac{f(\zeta)}{\zeta - z}\,d\zeta.$$

Now if $|\zeta - z_0| = r$, then

$$\frac{1}{\zeta - z} = \frac{1}{(\zeta - z_0) - (z - z_0)}$$

$$= \frac{1}{\zeta - z_0} \frac{1}{1 - \dfrac{z - z_0}{\zeta - z_0}}$$

$$= \sum_{n=0}^{\infty} \frac{(z - z_0)^n}{(\zeta - z_0)^{n+1}}$$

Hence

$$\frac{f(\zeta)}{\zeta - z} = \sum_{n=0}^{\infty} g_n(\zeta),$$

where

$$g_n(\zeta) = \frac{f(\zeta)(z - z_0)^n}{(\zeta - z_0)^{n+1}}.$$

Since $C(z_0, r)^*$ is compact, $|f|$ is bounded on this set: let $M = \sup\{|f(\zeta)| : |\zeta - z_0| = r\}$. Then $|g_n(\zeta)| \leqslant (M/r)(\rho/r)^n$ for $|\zeta - z_0| = r$, so $\sum g_n$ is uniformly convergent on $C(z_0, r)^*$, by 1.2.4 (the 'M-test'). Hence

$$\int_{C(z_0, r)} \left(\sum_{n=0}^{\infty} g_n\right) = \sum_{n=0}^{\infty} \int_{C(z_0, r)} g_n,$$

so that

$$f(z) = \frac{1}{2\pi i} \sum_{n=0}^{\infty} \int_{C(z_0, r)} g_n$$

$$= \sum_{n=0}^{\infty} a_n (z - z_0)^n,$$

where

$$a_n = \frac{1}{2\pi i} \int_{C(z_0, r)} \frac{f(z)}{(\zeta - z_0)^{n+1}} d\zeta. \tag{1}$$

Uniqueness of the a_n follows from 1.2.8 or 1.7.8. By 2.1.4, the value of the integral in (1) is the same for all r in $(0, R)$.

The series defined in the theorem is called the **Taylor series** for f at z_0. By 1.2.8, if we have obtained by any method (or as the definition) a power series $\sum b_n(z-z_0)^n$ that converges to f on some disc $D(z_0, R')$ (where, possibly, $R' < R$), then this series is the Taylor series for f at z_0. Because of this, it is by no means always necessary to evaluate the integrals in order to find the a_n. Furthermore, 2.2.1 tells us that $\sum b_n(z-z_0)^n$ converges whenever $|z-z_0| < R$, a fact that is not always trivial from consideration of the series itself.

As an illustration, if we are asked to find the Taylor series for exp at 1, we simply notice that

$$e^z = e\,.\,e^{z-1} = \sum_{n=0}^{\infty} \frac{e}{n!}(z-1)^n.$$

Example. If a function is given as the ratio of two series, we can use 1.2.6 to obtain a recurrence relation for the coefficients in its Taylor series. For example, consider the function tan at 0. Since the function is odd, the Taylor series will be of the form $a_1 z + a_3 z^3 + a_5 z^5 + \cdots$ (cf. exercise 5 of 1.2). Hence we have

$$z - \frac{z^3}{3!} + \frac{z^5}{5!} - \cdots = \left(1 - \frac{z^2}{2!} + \frac{z^4}{4!} - \cdots\right)(a_1 z + a_3 z^3 + \cdots)$$

on a disc centre 0. Applying 1.2.6 to the product on the right-hand side, and equating coefficients of powers of z, we obtain $a_1 = 1$, and

$$a_{2n+1} - \frac{a_{2n-1}}{2!} + \frac{a_{2n-3}}{4!} - \cdots + \frac{(-1)^n}{(2n)!} a_1 = \frac{(-1)^n}{(2n+1)!}.$$

Since tan is differentiable on $D(0, \pi/2)$, the series converges on this disc.

Higher derivatives

As an easy deduction from the preceding theorem, we have a striking result that would be totally false if stated for real functions:

2.2.2. *If G is an open subset of $\mathbf{C}$, and f is a differentiable function from G to $\mathbf{C}$, then f has all derivatives on G.*

Proof. Choose a point z of G. Since G is open, there exists $\rho > 0$ such that $D(z, \rho) \subseteq G$. By 2.2.1, there exist a_n such that $f(\zeta) = \sum_{n=0}^{\infty} a_n(\zeta - z)^n$ for $|\zeta - z| < \rho$. By 1.5.7, it follows that f has all derivatives at z.

More exactly, we have $f^{(n)}(z) = n!a_n$ for each n. Hence the coefficients in the Taylor series can be expressed, not only as integrals, but also as derivatives. If we equate the two expressions, we obtain

$$f^{(n)}(z) = \frac{n!}{2\pi i} \int_{C(z, \rho')} \frac{f(\zeta)}{(\zeta - z)^{n+1}} \, d\zeta,$$

where $0 < \rho' < \rho$. Suppose now that f is differentiable on $D(z_0, R)$, and that $|z - z_0| < R$. Take r such that $|z - z_0| < r < R$, and ρ' such that $0 < \rho' < r - |z - z_0|$. Then 2.1.4 shows that

$$\int_{C(z, \rho')} \frac{f(\zeta)}{(\zeta - z)^{n+1}} \, d\zeta = \int_{C(z_0, r)} \frac{f(\zeta)}{(\zeta - z)^{n+1}} \, d\zeta.$$

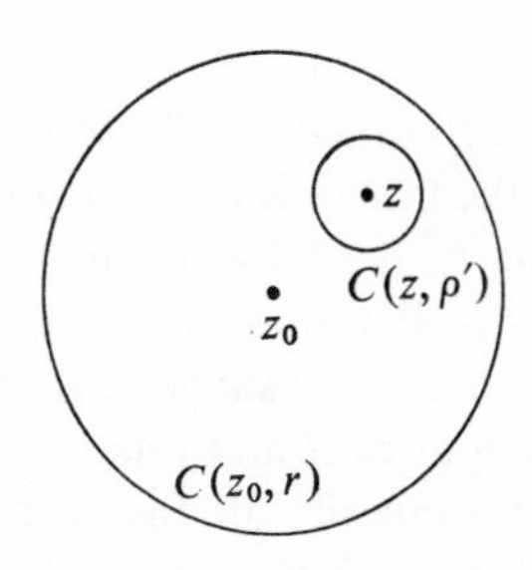

Hence we have established the following result, which gives an expression for higher derivatives analogous to Cauchy's integral formula:

2.2.3. *Suppose that f is differentiable on $D(z_0, R)$, and that $0 < r < R$. Then, for all z in $D(z_0, r)$,*

$$f^{(n)}(z) = \frac{n!}{2\pi i} \int_{C(z_0, r)} \frac{f(\zeta)}{(\zeta - z)^{n+1}} \, d\zeta \qquad (n = 0, 1, 2, \ldots).$$

Notice that this formula is obtained if one differentiates Cauchy's integral formula under the integral sign with respect to z. A direct justification of this process is given in many books, but, as the above shows, this is superfluous to our development.

Example. (Cf. the example following 2.1.7.) To evaluate

$$\int_{C(0,\,1)} \left(\frac{z-a}{z-b}\right)^2 dz,$$

where $|b| < 1$. Let $f(z) = (z-a)^2$. The integral is $2\pi i f'(b) = 4\pi i(b-a)$.

We draw the reader's attention to the fact that 2.2.1 (apart from the formula for a_n) and 2.2.2 can be stated without any reference to integration. These results were known for many years before any proof avoiding integration was devised, and those that have been devised are distinctly more laborious than the method (using integration) that we have given here. (See Whyburn, *Topological Analysis*.)

The order of a zero

If f is differentiable and not identically zero on a neighbourhood of z_0, and $f(z_0) = 0$, then there is a unique integer p such that the Taylor series for f at z_0 if of the form $\sum_{n=p}^{\infty} a_n(z-z_0)^n$, where $a_p \neq 0$. The function f is then said to have a **zero** of **order** p at z_0. A zero of order 1 is also called a **simple** zero. Since $n!\,a_n = f^{(n)}(z_0)$, f has a zero of order p at z_0 if and only if $f^{(r)}(z_0) = 0$ for $r < p$ and $f^{(p)}(z_0) \neq 0$. For example, all the zeros of sin are simple, since its derivative is cos, which is always non-zero when sin is zero.

For the moment, we will say no more about the orders of zeros, but we shall return to the topic in 2.5.

Finite Taylor expansions

If $f(z) = \sum_{n=0}^{\infty} a_n(z-z_0)^n$ for $|z-z_0| < R$, then, for any positive integer p, we can write

$$f(z) = \sum_{n=0}^{p-1} a_n(z-z_0)^n + (z-z_0)^p g_p(z), \tag{1}$$

where

$$g_p(z) = a_p + a_{p+1}(z - z_0) + \cdots.$$

Clearly, g_p is differentiable on $D(z_0, R)$, and

$$g_p(z_0) = a_p \tag{2}$$

Expressed differently, the function g_p defined on $D(z_0, R)$ by (1) and (2) is differentiable. The case $p = 1$ is often useful. We restate it in a slightly different form:

2.2.4. *Suppose that f is differentiable on an open set G. If $z_0 \in G$, and*

$$g(z) = \frac{f(z) - f(z_0)}{z - z_0} \qquad (z \neq z_0), \qquad g(z_0) = f'(z_0),$$

then g is differentiable on G.

Proof. The function g defined in this way is clearly differentiable on $G \backslash \{z_0\}$, and the above remarks show that it is also differentiable at z_0.

Consequences of the differentiability of derivatives

Recall the first statement in 2.1.3: every differentiable function on a star-shaped, open set is itself a derivative. We have now proved the (far more unexpected) converse: every derivative on an open set is itself differentiable. Using this, we can prove the following result (known as Morera's theorem), which is essentially the converse of the second statement in 2.1.3.

2.2.5. *Let G be an open subset of $\mathbf{C}$, and let f be a continuous function from G to $\mathbf{C}$ such that $\int_{\partial T} f = 0$ for every triangle T with $\operatorname{co}(T) \subseteq G$. Then f is differentiable on G.*

Proof. Take z in G. There exists $r > 0$ such that $D(z, r) \subseteq G$. By 2.1.2, f is the derivative of another function on $D(z, r)$. Hence f is differentiable at z.

Another application states, loosely speaking, that a non-zero differentiable function has a differentiable logarithm:

2.2.6. *Suppose that f is differentiable and non-zero on a star-shaped open set G. Then there is a differentiable function h such that $e^{h(z)} = f(z)$ $(z \in G)$.*

Proof. Since f'/f is differentiable on G, 2.1.3 shows that there is a function h such that $h' = f'/f$ on G. By adding a suitable constant, we may suppose that, for some a in G, $h(a)$ is a logarithm of $f(a)$. Now

$$\frac{d}{dz}(f(z)e^{-h(z)}) = (f'(z) - h'(z)f(z))e^{-h(z)} = 0 \qquad (z \in G),$$

so $f(z)e^{-h(z)}$ is constant on G. Since $f(a)e^{-h(a)} = 1$, the constant is 1.

Points at which a particular value is attained

Using the full strength of the uniqueness theorem for power series, we can prove a similar result for differentiable functions:

2.2.7 Theorem. *Let G be a connected open subset of $\mathbf{C}$, and let f and g be differentiable functions from G to $\mathbf{C}$. Suppose that there exist points z_n of G $(n = 0, 1, 2, \ldots)$ such that* (i) *$z_n \neq z_0$ for $n \geqslant 1$,* (ii) *$z_n \to z_0$ as $n \to \infty$,* (iii) *$f(z_n) = g(z_n)$ for $n \geqslant 1$. Then $f(z) = g(z)$ $(z \in G)$.*

Proof. By 1.2.8, the Taylor series for f and g at z_0 coincide. Let H be the set of points z of G such that the Taylor series for f and g at z coincide, i.e. such that $f^{(n)}(z) = g^{(n)}(z)$ for all n. Since $f^{(n)}$ and $g^{(n)}$ are continuous for each n, H is closed in G. If $z \in H$, then there exists $r > 0$ such that $f = g$ on $D(z, r)$. For ζ in $D(z, r)$, we then have $f = g$ on a neighbourhood of ζ, from which it follows that $\zeta \in H$. Hence H is also open. Since G is connected, it now follows that $H = G$, so that $f(z) = g(z)$ $(z \in G)$.

2.2.8 Corollary. *Let G be a connected open subset of $\mathbf{C}$, and let f be a differentiable function from G to $\mathbf{C}$ that is not constant on G. Then, given a in G and λ in C, there exists $r(a) > 0$ such that $f(z) \neq \lambda$ whenever $0 < |z - a| < r(a)$.*

Proof. Suppose that the statement is false. Then there exist points a, a_n of G $(n \geqslant 1)$ and λ in $\mathbf{C}$ such that (i) $a_n \neq a$ for all n, (ii) $a_n \to a$, and (iii) $f(a_n) = \lambda$ for all n. But then 2.2.7 shows that $f(z) = \lambda$ for all z in G.

The case where $\lambda \neq f(a)$ in 2.2.8 can be proved using only the continuity of f. It is the case where $\lambda = f(a)$ that is of interest: 2.2.8 then asserts that there is a neighbourhood of a on which f does not again assume the value λ. The reader may care to show by an example that this statement is not true for real functions.

2.2.9 Corollary. *Let G be a connected open subset of $\mathbf{C}$, and let f be a non-constant, differentiable function from G to $\mathbf{C}$. Let A be a bounded, closed subset of G. Then, for each λ in $\mathbf{C}$, $\{z \in A : f(z) = \lambda\}$ is finite.*

Proof. Let $r(a)$ be as in 2.2.8. Since A is compact, there is a finite set $a_1, \ldots, a_n$ such that

$$A \subseteq \bigcup_{j=1}^{n} D(a_j, r(a_j)).$$

If $z \in A$ and $f(z) = \lambda$, then z must be one of the a_j.

Notice that the conclusion of 2.2.9 can be false if we drop the assumption that A is closed. For example, let $f(z) = \sin(1/z)$, and let $A = D(1, 1)$. Then f is zero at each of the points $1/n\pi$ of A.

The limit of a sequence of differentiable functions

2.2.10 Theorem. *Let G be an open subset of $\mathbf{C}$, and let $\{f_n\}$ be a sequence of differentiable functions from G to $\mathbf{C}$. Suppose that $f_n(z) \to f(z)$ $(z \in G)$, and that convergence is uniform on each compact subset of G. Then f is differentiable on G, and $f'_n \to f'$ uniformly on each compact subset of G.*

Proof. Choose a point a of G. There exists $R > 0$ such that $D(a, R) \subseteq G$. Let T be a triangle in $D(a, R)$. Then $\int_{\partial T} f_n = 0$ for each n, and $f_n \to f$ uniformly on ∂T^*, so $\int_{\partial T} f = 0$. Hence, by 2.2.5, f is differentiable on $D(a, R)$ (and, in particular, at a).

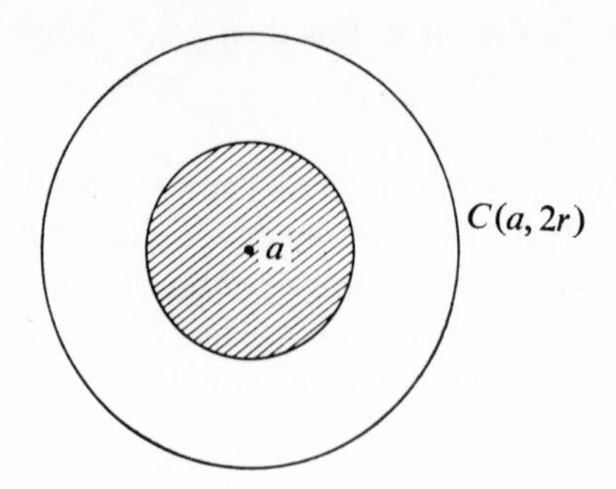

Now take $\epsilon > 0$ and r such that $0 < r < R/2$. By 2.2.3, for all z such that $|z-a| \leqslant r$,

$$f'_n(z) - f'(z) = \frac{1}{2\pi i}\int_{C(a,\,2r)} \frac{f_n(\zeta) - f(\zeta)}{(\zeta - z)^2}\,d\zeta.$$

There exists N such that for $n \geqslant N$, $|f_n(\zeta) - f(\zeta)| \leqslant \epsilon r$ whenever $|\zeta - a| = 2r$. For such n, we have

$$|f'_n(z) - f'(z)| \leqslant \frac{1}{2\pi}\frac{\epsilon r}{r^2}4\pi r = 2\epsilon.$$

Hence $f'_n \to f'$ uniformly on $\{z : |z-a| \leqslant r\}$.

Let A be a compact subset of G. The above shows that, for each a in A, there exists $r > 0$ such that $f'_n \to f'$ uniformly on $\{z\colon |z-a| \leqslant r\}$. A finite number of such sets covers A, from which it follows that $f'_n \to f'$ uniformly on A.

The theorem on differentiation of power series (1.5.7) is, of course, a special case of this result.

The example $f_n(x) = (1/n)\sin nx$ shows that 2.2.10 is not true for real functions.

Exercises 2.2

1 Find the Taylor series for cos and sin at $\pi/4$ (i) by using the addition formulae for cos and sin, (ii) by differentiating to find the coefficients.

2 Show that the Taylor series for $1/(1-z+z^2)$ at 0 is $\sum_{n=0}^{\infty} a_n z^n$, where $a_0 = a_1 = 1$, $a_2 = 0$, and $a_{n+3} = -a_n$ $(n \geqslant 0)$. What is the radius of convergence?

3 Evaluate $\int_{C(i,\,2)} e^z/(z-1)^n \, dz$, where n is a positive integer.

4 Find the order of the zero of each of the following functions at 1: $e^{z-1}-1$, $z \sin z$, $z^5 - 3z^4 + 8z^2 - 9z + 3$.

5 By considering the function $x \mapsto x|x|$ (or otherwise), show that 2.2.4 is false for real functions.

6 Would it have been logically possible to leave out the proof of 1.5.7 because this result is a special case of 2.2.10?

7 Let f and g be differentiable functions on a connected open set G, and suppose that $f(z)g(z) = 0$ $(z \in G)$. Prove that either f or g is identically zero on G.

8 Let G be a connected open set that contains $\bar{z}$ whenever it contains z. If f is differentiable on G, and

$$\overline{g(z)} = f(\bar{z}) \qquad (z \in G),$$

show that g is differentiable on G. If f is real-valued on $\mathbf{R} \cap G$, deduce that

$$f(\bar{z}) = \overline{f(z)} \qquad (z \in G).$$

9 Suppose that f is a non-constant entire function. Show that, for each λ in $\mathbf{C}$, the set $\{z \in \mathbf{C} : f(z) = \lambda\}$ is finite or countable. Deduce that $f(\mathbf{C})$ is uncountable.

10 Let f be differentiable on an open set G, and suppose that f has a finite number of zeros in G. Show that f can be written in the form pq, where p is a polynomial, and q is non-zero and differentiable throughout G.

2.3. Entire functions and polynomials

Recall that an entire function is a function from $\mathbf{C}$ to $\mathbf{C}$ that is differentiable on the whole of $\mathbf{C}$. By 2.2.1, these are simply the functions that are the sums of power series that converge for all complex arguments. As examples, we may mention polynomials, exp, sin, cos, and products and compositions of these.

2.3.1 Theorem (Liouville). *If f is an entire function, and $f(z)/z \to 0$ as $z \to \infty$ (in particular, if f is bounded), then f is constant.*

Proof. There exist a_n such that $f(z) = \sum_{n=0}^{\infty} a_n z^n$ $(z \in \mathbf{C})$. Let $M(r) = \sup\{|f(z)| : |z| = r\}$. By 1.7.9, $|a_n| \leqslant M(r)/r^n$ for all n and $r > 0$. But, by hypothesis, $M(r)/r \to 0$ as $r \to \infty$. Hence $a_n = 0$ for all $n \geqslant 1$, and $f(z) = a_0$ for all z.

More generally, the same reasoning gives:

2.3.2. *Let f be an entire function, and suppose that there exist $k > 0$, $R > 0$ and a positive integer n such that $|f(z)| \leqslant k|z|^n$ whenever $|z| > R$. Then f is a polynomial of degree not greater than n.*

Proof. Let $f(z) = \sum_{n-0}^{\infty} a_n z^n$, and take $m > n$. For $r > R$, $M(r) \leqslant kr^n$, so $|a_m| \leqslant kr^{n-m}$. Hence $a_m = 0$.

Conversely, the value of a polynomial approximates to its leading term for large $|z|$, in the following sense:

2.3.3. *Let $p(z) = a_0 + a_1 z + \cdots + a_n z^n$, where $a_n \neq 0$. Then $p(z)/a_n z^n \to 1$ as $z \to \infty$.*

Proof. For $z \neq 0$,

$$\frac{p(z)}{a_n z^n} - 1 = \frac{b_0}{z^n} + \frac{b_1}{z^{n-1}} + \cdots + \frac{b_{n-1}}{z},$$

where $b_j = a_j/a_n$ for each j. Given $\epsilon > 0$, there exists $\delta > 0$ such that for $|\zeta| \leqslant \delta$,

$$|b_0 \zeta^n + b_1 \zeta^{n-1} + \cdots + b_{n-1} \zeta| \leqslant \epsilon.$$

Then

$$\left|\frac{p(z)}{a_n z^n} - 1\right| \leqslant \epsilon \qquad \text{for } |z| \geqslant \frac{1}{\delta}.$$

Hence for sufficiently large $|z|$, we have $|p(z)| > \frac{1}{2}|a_n z^n|$. If $n \geqslant 1$, it follows that $1/p(z) \to 0$ as $z \to \infty$. We can now prove the theorem on polynomials that was promised in section 1.1.

2.3.4 Theorem. *Every non-constant complex polynomial has zeros.*

Proof. If p is a polynomial without zeros, then $1/p$ is an entire function. By 2.3.3, $1/zp(z) \to 0$ as $z \to \infty$, so, by 2.3.1, $1/p$ is constant. Hence p is constant.

Notice that we have used the analysis of both differentiation and integration in proving this purely algebraic result.

At this point, we make a brief excursion into algebra to show how the existence of zeros implies that every complex polynomial is a product of 'linear' factors, that is, of polynomials of degree 1. A special case of the 'division algorithm' states that if p is a polynomial of degree $n \geqslant 1$, and $a \in \mathbf{C}$, then there exist a polynomial q of degree $n-1$ and b in $\mathbf{C}$ such that $p(z) = (z-a)q(z) + b\ (z \in \mathbf{C})$. The proof (by induction on n) is elementary. We are interested here in the following corollary: if $p(a) = 0$, then there is a polynomial q of degree $n-1$ such that $p(z) = (z-a)q(z)\ (z \in \mathbf{C})$. One consequence of this is that a polynomial of degree n cannot have more than n zeros, and therefore that if two polynomials of degree n (or less than n) have the same value at $n+1$ different points, then they are identical.

2.3.5 Corollary. *If p is a complex polynomial of degree n, then there exist unique complex numbers $c, a_1, \ldots, a_n$ such that*

$$p(z) = c(z-a_1)\cdots(z-a_n).$$

Proof. We prove both existence and uniqueness by induction on the degree of the polynomial. Suppose that every polynomial of degree $n-1$ has linear factors, and let p be a polynomial of degree n. By 2.3.4, p has a zero, say a. By the division algorithm, there is a polynomial q of degree $n-1$ such that $p(z) = (z-a)q(z)\ (z \in \mathbf{C})$. Hence p has linear factors.

To prove uniqueness, it helps to introduce an ordering into $\mathbf{C}$.

Write $x+iy \,[\, x'+iy'$ to mean: $x<x'$, or $x=x'$ and $y \leqslant y'$. Suppose, now, that uniqueness has been established for polynomials of degree $n-1$, and that

$$p(z)=c(z-a_1)\cdots(z-a_n)=d(z-b_1)\cdots(z-b_n),$$

where $c,\ d \neq 0$ and $a_j \,[\, a_{j+1},\ b_j \,[\, b_{j+1}$ for each j. Considering the coefficient of z^n, we have $c=d$. Now a_1 and b_1 are both the zero of p that comes first in the ordering [, so $a_1=b_1$. Therefore $(z-a_2)\cdots(z-a_n)=(z-b_2)\cdots(z-b_n)$ for all $z \neq a_1$, and hence (by the remark preceding the theorem or by continuity) for all z. The induction hypothesis now shows that $a_j=b_j$ for $j=2,\ldots,n$.

We now have one of the classic applications of the complex numbers to the understanding of the real numbers. Neat as the theory of complex functions may be, it is perhaps permissible to object to it on the grounds that the complex numbers are an unnatural creation, whereas the definition of the real numbers (though, like all Mathematics, an invention of the mind) is at least motivated by our experience of the physical world. But even if this viewpoint is taken, the construction of the complex numbers must be accepted as justified if it enables us to prove results about the reals that cannot easily be proved otherwise. The following theorem on factorization of real polynomials is an example of such a result (further application of complex function theory to problems concerning real functions are given in 3.1, 3.2 and 3.3).

2.3.6 Corollary. *Every real polynomial can be expressed as a product of* (i) *linear real polynomials, and* (ii) *quadratic real polynomials of the form* $(x-\lambda)^2+\mu^2$, *where* $\mu>0$.

Proof. Let p be a real polynomial. We can evaluate $p(\alpha)$ for any complex α, and it is clear that $p(\bar{\alpha})=\overline{p(\alpha)}$. Hence if $p(\alpha)=0$, then $p(\bar{\alpha})=0$, so $p(x)$ is a product of factors of the form $(x-\alpha)$ (α real) or of the form $(x-\alpha)(x-\bar{\alpha})$ (α complex). The result follows, for if $\alpha=\lambda+i\mu$, then

$$(x-\alpha)(x-\bar{\alpha})=(x-\lambda)^2+\mu^2.$$

Exercises 2.3

1 Think of an example of an entire function that is not a polynomial and (i) has no zeros, (ii) has exactly one zero, (iii) has an infinite number of zeros.

2 If f is a function such that $f'(z) = f(z)$ for all z in $\mathbf{C}$, show that f is a constant multiple of exp.

3 Prove (i) directly from the intermediate value theorem, and (ii) using the results of this section, that every real polynomial of odd degree has a real zero.

4 Let p be a non-constant polynomial. If $\alpha > 0$ is given, show that there exists $\beta > 0$ such that $\{p(z): |z| > \alpha\}$ contains $\{z: |z| > \beta\}$.

5 Using only 2.3.3 and compactness, show that for any polynomial p, there is a point z_0 such that $|p(z_0)| = \inf\{|p(z)|: z \in \mathbf{C}\}$.

6 Suppose that f is an entire function, and that there exist a in $\mathbf{C}$ and $\epsilon > 0$ such that $|f(z) - a| > \epsilon$ for all z. Show that f is constant. Deduce that if f is a non-constant entire function, then $f(\mathbf{C})$ is dense in $\mathbf{C}$. (A stronger result is proved later: see 2.5.9).

7 The following shows how our results on polynomials, together with a little linear algebra, enable us to prove the assertion about multiplicative structures on $\mathbf{R}^n$ made in 1.1. Instead of assuming that the non-zero elements form a multiplicative group, it is sufficient to assume that there are no *zero-divisors*, i.e. that $ab \neq 0$ whenever $a \neq 0$ and $b \neq 0$. The proof outlined here is a combination of the methods of Dickson (*Linear Algebras*, 1914) and Palais (*American Mathematical Monthly*, 1968).

Suppose that an associative multiplication can be defined on $\mathbf{R}^n$ such that there are no zero-divisors, and such that for all a, b, c in $\mathbf{R}^n$ and λ in $\mathbf{R}$,

$$a(b+c) = ab + ac, \qquad (a+b)c = ac + bc,$$
$$(\lambda a)b = a(\lambda b) = \lambda(ab).$$

(Here $a+b$ and λa stand for the usual linear-space operations on $\mathbf{R}^n$, and ab denotes the multiplication; in the standard terminology, our hypotheses say that $\mathbf{R}^n$ is an *associative linear algebra* over $\mathbf{R}$.) Prove the following statements:

(i) If $a^2 = b^2$ and $ab = ba$, then $a = b$ or $a = -b$.

(ii) For a in $\mathbf{R}^n$, define L_a by: $L_a(x) = ax$ ($x \in \mathbf{R}^n$). If a is non-zero, then the range of L_a is the whole of $\mathbf{R}^n$. The non-zero elements of $\mathbf{R}^n$ form a group with respect to multiplication. In particular, there is a multiplicative identity, e.

(iii) For any x in $\mathbf{R}^n$, there exist $\lambda_0, \lambda_1, \ldots, \lambda_n$ in $\mathbf{R}$ such that $\lambda_0 e + \lambda_1 x + \cdots + \lambda_n x^n = 0$.

(iv) For any x in $\mathbf{R}^n$, there exist μ_0, μ_1, μ_2 in $\mathbf{R}$ such that $\mu_0 e + \mu_1 x + \mu_2 x^2 = 0$. (Factorize the real polynomial in (iii).)

(v) If x is not a scalar multiple of e, then there exist α in $\mathbf{R}\backslash\{0\}$ and β in $\mathbf{R}$ such that $(\alpha x + \beta e)^2 = -e$.

(vi) Suppose, henceforth, that $n \geqslant 2$. There exists f such that $f^2 = -e$. Let

$$D^+ = \{x : xf = fx\}, \qquad D^- = \{x : xf = -fx\}.$$

Then D^+ is the linear subspace generated by e and f (use (v) and (i)). If the multiplication is commutative, then $n = 2$.

(vii) For any x,

$$x - fxf \in D^+, \qquad x + fxf \in D^-,$$

so $D^+ + D^- = \mathbf{R}^n$. If D^- contains a non-zero element a, then L_a maps D^+ into D^- and D^- into D^+, and hence $n = 4$.

The reader may also care to prove that $\mathbf{R}^n$, with the three given operations, is isomorphic to $\mathbf{C}$ in the case $n = 2$, and to the quaternions in the case $n = 4$. (For the latter, start by using the fact that a commutes with a^2 to show that if $a \in D^-\backslash\{0\}$, then a^2 is a negative real multiple of e.)

The hypothesis connecting multiplication with scalar multiplication can be replaced by the assumption that, for each a, the mappings $x \mapsto ax$ and $x \mapsto xa$ are continuous. For the distributive laws imply that

$$(\lambda a)\, b = a(\lambda b) = \lambda(ab)$$

for all rational λ, and continuity then enables us to deduce that the same relation holds for all real λ.

It has been shown that, even if associativity is abandoned, only one new case arises, viz. $n = 8$.

2.4. The modulus of a differentiable function

Maxima and minima

The next result – another straightforward deduction from Cauchy's integral formula – shows that the modulus of a non-constant, differentiable complex function does not have local maxima (cf. exercise 4 of 1.5).

2.4.1 Theorem. *Suppose that f is differentiable, not constant, on a neighbourhood of a. Then there exists, for each $R > 0$, a point z in $D(a, R)$ such that $|f(z)| > |f(a)|$.*

Proof. By 2.2.7, f is not constant on any $D(a, R)$. If $f(a) = 0$, the result follows at once. Suppose, then, that $f(a) \neq 0$, and that $|f(z)| \leqslant |f(a)|$ whenever $|z-a| < R$. Take r in $(0, R)$. By 2.1.7, we have

$$\frac{1}{2\pi}\int_0^{2\pi} \frac{f(a+r\,e^{it})}{f(a)}\,dt = 1.$$

Write

$$\frac{f(a+r\,e^{it})}{f(a)} = u(t)+iv(t).$$

Then $\int_0^{2\pi}(u+iv) = 2\pi$, so $\int_0^{2\pi} u = 2\pi$ and $\int_0^{2\pi} v = 0$. Now $|u(t)+iv(t)| \leqslant 1$ for $0 \leqslant t \leqslant 2\pi$. If $u(t_0) < 1$ for some t_0 in $[0, 2\pi]$, then, by continuity, there exist $\alpha < 1$ and t_1, t_2 such that $0 \leqslant t_1 < t_2 \leqslant 2\pi$ and $u(t) \leqslant \alpha$ for $t_1 \leqslant t \leqslant t_2$. From this it follows that $\int_0^{2\pi} u < 2\pi$, which is a contradiction. Hence $u(t) = 1$ and $v(t) = 0$ for $0 \leqslant t \leqslant 2\pi$. In other words, $f(a+re^{it})$ is equal to the constant value $f(a)$ for $0 \leqslant t \leqslant 2\pi$ and $0 < r < R$.

The rest of this section consists of applications of this theorem. First we note:

2.4.2 Corollary. *If f is differentiable, not constant, on $D(a,R)$, and $f(a) \neq 0$, then there exists a point z in $D(a,R)$ such that $|f(z)| < |f(a)|$.*

Proof. There exists r such that $0 < r \leqslant R$ and $f(z) \neq 0$ for $|z-a| < r$. The function $1/f$ is differentiable on $D(a,r)$. The result follows by applying 2.4.1 to $1/f$.

The next corollary (sometimes called the 'maximum modulus principle') is a typical example of the deduction of a 'global' result from a 'local' one.

2.4.3 Corollary. *Let f be differentiable on an open subset G of $\mathbf{C}$, and let A be a bounded, closed subset of G. Then* $\sup\{|f(z)| : z \in A\}$ *is attained at a boundary point of A.*

Proof. Since A is compact, $\sup\{|f(z)| : z \in A\}$ is attained at a point of A, by 0.9. The result is trivial if f is constant. Otherwise, 2.4.1 shows that the supremum is not attained at any interior point of A.

Mappings of the unit disc

Given two open subsets A, B of $\mathbf{C}$, the following question presents itself: what differentiable mappings are there taking A onto B? At first sight, this question may seem hopeless, but with the aid of the theory we have developed, some surprisingly powerful answers can be given. For instance, the results of section 2.3 tell us that, unless B is dense in $\mathbf{C}$, there are no differentiable functions mapping $\mathbf{C}$ on to B. The question can be made more restrictive by looking for differentiable functions that are one-to-one and have a differentiable inverse (such functions are said to be **conformal**).

The following sequence of results gives some information on these questions in the case when both A and B are the unit disc $D(0,1)$ (which, for the rest of this section, we denote simply by D). Though rather special, the results are of considerable importance in some branches of complex function theory.

2.4.4 (Schwarz's lemma). *Suppose that f is a differentiable function mapping D into D, and that $f(0) = 0$. Then:*

either (i) $f(z) = cz \quad (z \in D)$ *for some c with $|c| = 1$,*

or (ii) $|f(z)| < |z|$ *whenever* $0 < |z| < 1$.

Proof. By 2.2.4, there is a differentiable function g on D such that $f(z) = zg(z)$ $(z \in D)$. Take r in $(0,1)$. Then

$$\begin{aligned} 1 &\geqslant \sup\{|f(z)| : |z| = r\} \\ &= r\sup\{|g(z)| : |z| = r\} \\ &= r\sup\{|g(z)| : |z| \leqslant r\}, \qquad \text{by 2.4.3.} \end{aligned}$$

Take z in D. For each r such that $|z| < r < 1$, we have just shown that $|g(z)| \leqslant 1/r$. Hence $|g(z)| \leqslant 1$.

If there is a point z_1 in D such that $|g(z_1)| = 1$, then $|g|$ has a local maximum at z_1, so, by 2.4.1, $g(z) = g(z_1)$ and $f(z) = zg(z_1)$ for all z in D. Otherwise, $|f(z)| < |z|$ whenever $0 < |z| < 1$.

2.4.5 Corollary. *If f is a differentiable mapping of D onto D, with a differentiable inverse, and $f(0) = 0$, then $f(z) = cz$ $(z \in D)$ for some c with $|c| = 1$.*

Proof. Applying 2.4.4 to f and to its inverse, we have $|f(z)| \leqslant |z|$ and $|z| \leqslant |f(z)|$ for z in D. Hence $|f(z)| = |z|$ $(z \in D)$, and alternative (i) of 2.4.4 must hold.

For a in D, define g_a by:

$$g_a(z) = \frac{z-a}{1-\bar{a}z} \qquad (|z| < 1/|a|).$$

Then g_a is clearly differentiable on its domain. To show that g_a maps D into D was exercise 3 of 1.1. It is interesting to note that an alternative method is afforded by the results of the present section. We show that $|g_a(z)| = 1$ when $|z| = 1$; since g_a is not constant, it follows, by 2.4.3 and 2.4.1, that $|g_a(z)| < 1$ when $|z| < 1$. If $|z| = 1$, then $|1 - \bar{a}z| = |1 - a\bar{z}| = |z - a|$: the first equality is obtained on taking the conjugate, and the second on multiplying by z. This gives the desired result.

2.4.6. *For each a in D, g_a is a differentiable mapping of D onto D, with differentiable inverse g_{-a}. Every conformal mapping of D onto D is of the form cg_a, where $a \in D$ and $|c| = 1$.*

Proof. Take w in D. Then $g_a(z) = w$ if and only if $w - \bar{a}zw = z - a$, or

$$z = \frac{w + a}{1 + \bar{a}w} = g_{-a}(w).$$

Let f be a conformal mapping of D onto D, and let a be the unique point of D such that $f(a) = 0$. Then $f \circ g_{-a}$ is a conformal mapping of D onto D, and maps 0 to 0. Therefore, by 2.4.5, there is a complex number c with unit modulus such that

$$(f \circ g_{-a})(\zeta) = c\zeta \qquad (\zeta \in D).$$

Applying this with $\zeta = g_a(z)$, we have $f(z) = cg_a(z) \quad (z \in D)$.

Various results on differentiable and conformal mappings can be deduced from 2.4.4 and 2.4.5 by considering suitable compositions of functions as in the proof of 2.4.6. Further examples are given in exercises 3 and 4.

Exercises 2.4

1 Let f be an entire function that has no zeros, and for $r > 0$, let $m(r) = \inf\{|f(z)| : |z| = r\}$. Show that m is a non-increasing function.

2 If f is differentiable on $D(a, 1)$, and $|f(z) - f(a)| \leqslant k$ for z in $D(a, 1)$, show that $|f(z) - f(a)| \leqslant k|z - a|$ for z in $D(a, 1)$.

3 If f is a differentiable mapping of D into D, with $f(a) = 0$, show that $|f(z)| \leqslant |g_a(z)|$ $(z \in D)$, where g_a is defined as in 2.4.6.

4 Let $V = \{z : \operatorname{Im} z > 0\}$. For a in V, let

$$h_a(z) = \frac{z - a}{z - \bar{a}} \qquad (z \neq \bar{a}).$$

Prove that h_a is a conformal mapping of V onto D. (What happens to the modulus of $h_a(z)$ when $\operatorname{Im} z > 0$ and when $\operatorname{Im} z \leqslant 0$?) By considering the function $h_a \circ f$, prove the following statements:

(i) If f is a differentiable mapping of D into V, with $f(0) = a$, then

$$|f(z)| \leqslant |a| \frac{1+|z|}{1-|z|} \qquad (z \in D).$$

(ii) If f is a conformal mapping of D onto V, with $f(0) = a$, then there is a number c with unit modulus such that

$$f(z) = \frac{ac - \bar{a}z}{c - z} \qquad (z \in D).$$

By considering the function $f \circ h_i^{-1}$, find a general expression for a conformal mapping f of V onto itself.

5 The following shows how one can prove that every non-constant complex polynomial has zeros without using integration. Check the details. Let p be a polynomial that is non-zero at z_0. We show that $|p|$ does not have a local minimum at z_0. Exercise 5 of 2.3 then gives the result. We can express p in the form

$$p(z) = a_0 + (z - z_0)^m [a_m + q(z)],$$

where a_0, $a_m \neq 0$ and $q(z) \to 0$ as $z \to z_0$. Write $a_0 = re^{i\alpha}$, $a_m = se^{i\beta}$, and let

$$z_\rho = z_0 + \rho e^{i(\alpha - \beta + \pi)/m}.$$

For sufficiently small $\rho > 0$, $|q(z_\rho)| \leqslant \frac{1}{2}s$, and then

$$|p(z_\rho)| \leqslant r - \tfrac{1}{2}\rho^m s < r.$$

By a slight modification to the definition of z_ρ, show also that $|p|$ does not have a local maximum at z_0, The argument applies to convergent power series as well as polynomials, and therefore gives an alternative proof of 2.4.1.

It is possible to prove that complex polynomials have zeros by applying purely algebraic techniques to the conclusion of exercise 3 of 2.3 (thereby avoiding use of the exponential function). The algebra involved is, however, quite sophisticated (see e.g. Lang, *Algebra*, pages 202–3).

2.5. Singularities; Laurent series

A function is said to have a **singularity** at z_0 if it is differentiable on a neighbourhood of z_0, except at z_0 itself. For example, the function $z \mapsto 1/z$ has a singularity at 0. The next stage in our investigation of complex functions is to examine the behaviour of functions near singularities. We find that there is an analogue of both Cauchy's integral formula and the Taylor series. These facts enable us to classify singularities. They also add greatly to our general understanding of complex functions – even of functions without singularities.

Recall that $D'(z_0, R)$ denotes $\{z : 0 < |z - z_0| < R\}$. The analogue of Cauchy's integral formula is the following result, in which the function is expressed as the difference between two integrals.

2.5.1. *Suppose that f is differentiable on $D'(z_0, R)$, and that $z \in D'(z_0, R)$. Let r_1, r_2 be such that*

$$0 < r_2 < |z - z_0| < r_1 < R.$$

Then

$$f(z) = \frac{1}{2\pi i}\int_{C(z_0, r_1)} \frac{f(\zeta)}{\zeta - z}\,d\zeta - \frac{1}{2\pi i}\int_{C(z_0, r_2)} \frac{f(\zeta)}{\zeta - z}\,d\zeta.$$

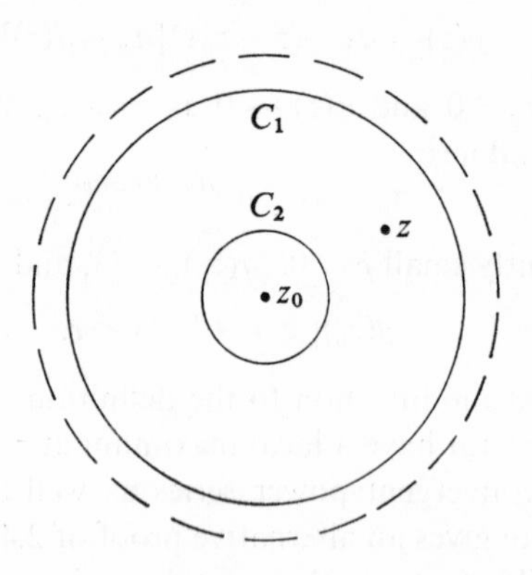

Proof. Write $C(z_0, r_j) = C_j$ $(j = 1, 2)$. For ζ in $D'(z_0, R)$, define

$$F(\zeta) = \frac{f(\zeta) - f(z)}{\zeta - z} \qquad (\zeta \neq z), \qquad F(z) = f'(z).$$

Then F is differentiable on $D'(z_0, R)$, by 2.2.4. Therefore, by 2.1.4, $\int_{C_1} F = \int_{C_2} F$. Now 2.1.5 shows that

$$\int_{C_1} \frac{1}{\zeta - z} d\zeta = 2\pi i,$$

and

$$\int_{C_2} \frac{1}{\zeta - z} d\zeta = 0.$$

Hence

$$\int_{C_1} F = \int_{C_1} \frac{f(\zeta)}{\zeta - z} d\zeta - 2\pi i f(z),$$

$$\int_{C_2} F = \int_{C_2} \frac{f(\zeta)}{\zeta - z} d\zeta.$$

The result follows.

The series expression for a function in a neighbourhood of a singularity z_0 takes the form of a power series of positive and negative powers of $z - z_0$. Before going on to the theorem on this, we give a precise meaning to such series and have a brief look at their properties.

Suppose that, for each integer n (positive and negative), a complex number a_n is given. If $\sum_{n=0}^{\infty} a_n = s_1$ and $\sum_{n=1}^{\infty} a_{-n} = s_2$, we write $\sum_{n=-\infty}^{\infty} a_n = s_1 + s_2$. It is easily verified that the following statement is equivalent: given $\epsilon > 0$, there exists N such that whenever $m, n > N$,

$$|(a_{-m} + \cdots + a_n) - (s_1 + s_2)| \leqslant \epsilon.$$

The next lemma shows how the properties of power series in negative powers can be deduced from corresponding properties of ordinary power series.

2.5.2 Lemma. *Suppose that* $\sum_{n=0}^{\infty} a_n z^n$ *is convergent to* $g(z)$ *for* $|z| < R$, *where* $R > 0$. *Then* $\sum_{n=0}^{\infty} a_n (z - z_0)^{-n}$ *is convergent (say to* $f(z)$*) for* $|z - z_0| > 1/R$, *and convergence is uniform on* $\{z: |z - z_0| \geqslant \rho\}$, *where* $\rho > 1/R$. *Also,* $f'(z) = -\sum_{n=1}^{\infty} n a_n (z - z_0)^{-n-1}$ *for* $|z - z_0| > 1/R$.

Proof. We have $f(z) = g[1/(z-z_0)]$ for $|z-z_0| > 1/R$. Take $\epsilon > 0$. By 1.2.5, there exists N such that $|g(z) - \sum_{n=0}^{p} a_n z^n| \leqslant \epsilon$ whenever $p \geqslant N$ and $|z| \leqslant 1/\rho$. Then $|f(z) - \sum_{n=0}^{p} a_n(z-z_0)^{-n}| \leqslant \epsilon$ whenever $p > N$ and $|z - z_0| \geqslant \rho$. Also, $g'(z) = \sum_{n=1}^{\infty} n a_n z^{n-1}$, by 1.5.7, so, by the composition rule,

$$f'(z) = -(z-z_0)^{-2} g'\left(\frac{1}{z-z_0}\right) = -\sum_{n=1}^{\infty} n a_n (z-z_0)^{-n-1}.$$

(Alternatively, the last statement follows from 2.2.10.)

2.5.3 Theorem. *Suppose that f is differentiable on $D'(z_0, R)$ where $R > 0$. Then there exist unique complex numbers a_n $(n = 0, \pm 1, \pm 2, \ldots)$ such that*

$$f(z) = \sum_{n=-\infty}^{\infty} a_n (z-z_0)^n$$

for z in $D'(z_0, R)$. If $0 < r < R$, then

$$a_n = \frac{1}{2\pi i} \int_{C(z_0, r)} \frac{f(\zeta)}{(\zeta - z_0)^{n+1}} d\zeta$$

for each n.

Proof. Take z in $D'(z_0, R)$, and r_1, r_2 such that $0 < r_2 < |z - z_0| < r_1 < R$.

Write $C(z_0, r_j) = C_j$ $(j = 1, 2)$. By 2.5.1,

$$f(z) = \frac{1}{2\pi i} \int_{C_1} \frac{f(\zeta)}{\zeta - z} d\zeta - \frac{1}{2\pi i} \int_{C_2} \frac{f(\zeta)}{\zeta - z} d\zeta.$$

As in 2.2.1,

$$\frac{1}{2\pi i} \int_{C_1} \frac{f(\zeta)}{\zeta - z} d\zeta = \sum_{n=0}^{\infty} a_n (z-z_0)^n,$$

where

$$a_n = \frac{1}{2\pi i} \int_{C_1} \frac{f(\zeta)}{(\zeta - z_0)^{n+1}} d\zeta \qquad (n = 0, 1, 2, \ldots).$$

We now deal with the other integral in a similar fashion. If $|\zeta - z_0| = r_2$, then

$$\frac{1}{\zeta - z} = -\frac{1}{z - z_0} \frac{1}{1 - \dfrac{\zeta - z_0}{z - z_0}} = -\sum_{n=1}^{\infty} \frac{(\zeta - z_0)^{n-1}}{(z - z_0)^n}.$$

By uniform convergence, exactly as in 2.2.1, it follows that

$$-\frac{1}{2\pi i}\int_{C_2}\frac{f(\zeta)}{\zeta-z}\,d\zeta=\sum_{n=1}^{\infty}a_{-n}(z-z_0)^{-n},$$

where

$$a_{-n}=\frac{1}{2\pi i}\int_{C_2}(\zeta-z_0)^{n-1}f(\zeta)\,d\zeta \qquad (n=1,2,\ldots).$$

By 2.1.4, the integrals for a_n can be taken round any circle with centre z_0 and radius less than R.

To prove uniqueness, suppose that $f(z)=\sum_{n=-\infty}^{\infty}b_n(z-z_0)^n$ for $0<|z-z_0|<R$ (it is sufficient, in fact, to assume this equality for $r_2<|z-z_0|<r_1$, where $0<r_2<r_1<R$). Choose r in $(0,R)$. Reasoning exactly as in the proof of 1.7.8, we see that

$$b_n=\frac{1}{2\pi i}\int_{C(z_0,\,r)}\frac{f(\zeta)}{(\zeta-z_0)^{n+1}}\,d\zeta$$

for each n.

The unique series found in 2.5.3 is called the **Laurent series** for f on $D'(z_0,R)$, or, briefly, the Laurent series for 'f at z_0'. The integral formulae for the coefficients in the Laurent series are the same as those for the Taylor series (and if f is differentiable at z_0, then the series are the same). On the other hand, of course, there is no possibility of expressing the Laurent coefficients in terms of derivatives of f at z_0. Exactly as in 1.7.9, we can prove the following result (known as *Cauchy's inequality*):

2.5.4. *Suppose that* $f(z)=\sum_{-\infty}^{\infty}a_n(z-z_0)^n$ *for* $0<|z-z_0|<R$, *and write* $M(r)=\sup\{|f(z)|:|z-z_0|=r\}$. *Then* $|a_n|\leqslant M(r)/r^n$ *for each* n.

The negative part of the Laurent series, i.e.

$$p(z)=\sum_{n=1}^{\infty}a_{-n}(z-z_0)^{-n},$$

is called the **principal part** of f at z_0. Since it is convergent for arbitrarily small $|z-z_0|$, it is convergent for all z different from z_0, giving a differentiable function on $\mathbf{C}\backslash\{z_0\}$. The function $f-p$ is differentiable wherever f is differentiable, and also at z_0 if it is given the value a_0 there, since

$$(f-p)(z) = \sum_{n=0}^{\infty} a_n(z-z_0)^n$$

for z in $D'(z_0, R)$.

Other forms of Laurent series

If f is differentiable on $\{z:|z| > R\}$, then g is differentiable on $D'(0, 1/R)$, where $g(z) = f(1/z)$. Applying 2.5.3 to g, we can deduce that there exist a_n such that

$$f(z) = \sum_{n=-\infty}^{\infty} a_n z^n \qquad (|z| > R).$$

Again the proof of 1.7.8 shows that the a_n are given by the formula of 2.5.3 (with $r > R$), and it follows that the inequality 2.5.4 is valid.

If we use the full strength of 2.1.4 (with the hypothesis sharpened by exercise 5 of 2.1), we obtain a variant of the Laurent theorem applicable to an annulus. The precise statement is as follows:

If f is differentiable on $\{z:r_1 < |z-z_0| < r_2\}$, then there exist unique complex numbers a_n such that

$$f(z) = \sum_{n=-\infty}^{\infty} a_n(z-z_0)^n$$

for $r_1 < |z-z_0| < r_2$. The a_n are given by the formula of 2.5.3, *with $r_1 < r < r_2$.*

The proof is identical to that of 2.5.1 and 2.5.3. We shall have little further use for Laurent series of this kind, but we give a simple example before leaving them. Consider the function

$$f(z) = \frac{1}{(z-1)(z-2)} = \frac{1}{z-2} - \frac{1}{z-1}$$

We can write down geometric series as follows:

$$\frac{1}{z-1} = -\sum_{n=0}^{\infty} z^n \quad (|z| < 1), \qquad \sum_{n=1}^{\infty} z^{-n} \quad (|z| > 1).$$

$$\frac{1}{z-2} = -\sum_{n=0}^{\infty} 2^{-n-1} z^n \quad (|z| < 2), \qquad \sum_{n=1}^{\infty} 2^{n-1} z^{-n} \quad (|z| > 2).$$

Combining the series as appropriate, we obtain the Laurent series for f on (i) $D(0,1)$, (ii) $\{z: 1 < |z| < 2\}$, (iii) $\{z: |z| > 2\}$.

Classification of singularities; the order of a function at a point

Let f have Laurent series $\sum_{n=-\infty}^{\infty} a_n(z-z_0)^n$ at z_0. If $a_n \neq 0$ for an infinite number of $n < 0$, we say that f has an **essential singularity** at z_0. Otherwise, the least integer n for which $a_n \neq 0$ is called the **order** of f at z_0. We denote this number by $\operatorname{ord}(f, z_0)$. Clearly, it is the unique integer n such that $f(z) = (z-z_0)^n f_1(z)$, where f_1 is differentiable and non-zero at z_0. If $\operatorname{ord}(f, z_0) = -n$, where $n > 0$, then f is said to have a **pole** of order n at z_0. A pole of order 1 is also called a **simple** pole. We now give some rules for combining functions of finite order:

2.5.5. *If f and g are of finite order at z_0, then:*
(i) $\operatorname{ord}(fg, z_0) = \operatorname{ord}(f, z_0) + \operatorname{ord}(g, z_0)$;
(ii) $\operatorname{ord}(1/f, z_0) = -\operatorname{ord}(f, z_0)$;
(iii) *if* $\operatorname{ord}(f, z_0) < \operatorname{ord}(g, z_0)$, *then*

$$\operatorname{ord}(f+g, z_0) = \operatorname{ord}(f, z_0).$$

Proof. (iii) is obvious on adding the Laurent series for f and g. To prove (i) and (ii), let $f(z) = (z-z_0)^m f_1(z)$, $g(z) = (z-z_0)^n g_1(z)$, where f_1, g_1 are differentiable and non-zero at z_0. Then

$$f(z)g(z) = (z-z_0)^{m+n} f_1(z) g_1(z),$$
$$1/f(z) = (z-z_0)^{-m}(1/f_1(z)),$$

from which the statements follow.

Somewhat similar rules apply to essential singularities:

2.5.6. *Suppose that f has an essential singularity at z_0, and that g is of finite order at z_0. Then fg and $f+g$ have essential singularities at z_0.*

Proof. If fg is of finite order at z_0, then so is f, by (i) and (ii) of 2.5.5. The second statement is obvious on adding the Laurent series for f and g.

Removable singularities. Throughout the above, we have assumed that $f(z)$ is defined for z close to z_0, but not necessarily at z_0 itself. However, if $\text{ord}(f, z_0) = n \geqslant 0$, and $f(z_0)$ was initially undefined, then f becomes differentiable at z_0 if we define $f(z_0)$ to be a_0. When this occurs, we say that f has a **removable singularity** at z_0. If $n > 0$, then of course f has a zero of order n at z_0. Examples of removable singularities have already arisen in 2.2.4 and in subtracting the principal part (see after 2.5.4). In future, we shall automatically assume that removable singularities have been removed. For instance, we shall speak of $f'(z_0)$ without always bothering to point out that this presupposes that $f(z_0)$ has been defined in the obvious way. If z_0 is a pole or essential singularity, then there is no point in assigning a value to $f(z_0)$, but for simplicity we shall make free use of (technically inaccurate) expressions like 'a function on G with singularities at $z_1, \ldots, z_n$'.

Examples. (i) Let $f(z) = 1/\sin z - 1/z$. At $n\pi$ $(n \neq 0)$, sin has a simple zero, so 1/sin has a simple pole, by 2.5.5 (ii). Hence f has a simple pole at these points, by 2.5.5 (iii). Now

$$f(z) = \frac{z - \sin z}{z \sin z},$$

and

$$z - \sin z = \frac{z^3}{3!} - \frac{z^5}{5!} + \cdots,$$

$$z \sin z = z^2 - \frac{z^4}{3!} + \cdots,$$

so $\text{ord}(z - \sin z, 0) = 3$ and $\text{ord}(z \sin z, 0) = 2$. Therefore $\text{ord}(f, 0) = 1$. In other words, f not only has a removable singularity at 0, but is even zero there.

(ii) The function $z \mapsto e^{1/z}$ has an essential singularity at 0, since its Laurent series there is $\sum_{n=0}^{\infty} 1/n!z^n$.

Values near a singularity

We now show how the different kinds of singularities (and also zeros) can be characterized in terms of the set of values assumed by the function around the point. The first result shows that the behaviour of a function near a pole or zero is quite similar to the behaviour of a polynomial for large $|z|$. The proof is also similar, as is apparent from a comparison with 2.3.2 and 2.3.3.

2.5.7. *Let f be differentiable in a neighbourhood of z_0, except possibly at z_0 itself. Then the following statement is equivalent to* $\operatorname{ord}(f, z_0) = n$: *there exist strictly positive numbers α. β, δ such that*

$$\alpha|z - z_0|^n \leqslant |f(z)| \leqslant \beta|z - z_0|^n$$

whenever $0 < |z - z_0| < \delta$.

Proof. Let the Laurent series for f at z_0 be $\sum a_n(z - z_0)^n$.

First, suppose that $\operatorname{ord}(f, z_0) = n$. Then $(z - z_0)^{-n} f(z) = a_n + g(z)$, where $g(z) \to 0$ as $z \to z_0$. Given $\epsilon > 0$, there exists $\delta > 0$ such that $|g(z)| \leqslant \epsilon|a_n|$ whenever $0 < |z - z_0| < \delta$. For such z,

$$(1 - \epsilon)\,|a_n| \leqslant |(z - z_0)^{-n} f(z)| = (1 + \epsilon)\,|a_n|.$$

Now suppose that the condition holds. If $m < n$, then, by 2.5.4, $|a_m| \leqslant M(r)/r^m \leqslant \beta r^{n-m}$ for $0 < r < \delta$, so $a_m = 0$. If also $a_n = 0$, then $(z - z_0)^{-n} f(z) \to 0$ as $z \to z_0$, contradicting the hypothesis.

In particular, f is differentiable at z_0 (or has a removable singularity there) if and only if it is bounded in a neighbourhood. Furthermore, 2.5.4 shows that it is sufficient if $(z - z_0) f(z) \to 0$ as $z \to z_0$.

With surprising ease, we can now deduce an impressive result about the values of a function near an essential singularity:

2.5.8 Theorem (Casorati-Weierstrass). *Suppose that f has an essential singularity at z_0. Then, given a in* $\mathbf{C}$, $\epsilon > 0$ *and* $\delta > 0$, *there exists* z

such that $|z-z_0|<\delta$ *and* $|f(z)-a|<\epsilon$. *In other words, in every neighbourhood of* z_0, *f comes arbitrarily close to all complex values.*

Proof. Suppose that there exist a in $\mathbf{C}$, $\epsilon>0$ and $\delta>0$ such that $|f(z)-a|\geqslant\epsilon$ whenever $0<|z-z_0|<\delta$. Let $g(z)=1/(f(z)-a)$. Then $|g(z)|\leqslant 1/\epsilon$ for z in $D'(z_0,\delta)$, so, by the remark above, g has a removable singularity at z_0. Now g is not identically zero, so $\operatorname{ord}(g,z_0)$ is a non-negative integer k. By 2.5.5, $\operatorname{ord}(f-a,z_0)=-k$. But this contradicts the assumption that f has an essential singularity at z_0.

2.5.9 Corollary. *Let f be an entire function that is not a polynomial. Then, given a in* $\mathbf{C}$, $\epsilon>0$ *and* $R>0$, *there exists z such that* $|z|>R$ *and* $|f(z)-a|<\epsilon$. *In other words, outside every bounded set, f comes arbitrarily close to all complex values.*

Proof. Let $g(z)=f(1/z)$. Then g has an essential singularity at 0. The result follows.

A theorem of Picard states that an entire function omits at most one value. For a proof, see Ahlfors [1], pages 297–8.

If a function does not have the properties described in either 2.5.7 or 2.5.8, then we must conclude that it is not differentiable on any 'punctured neighbourhood' $D'(z_0,r)$ of z_0. By way of illustration, consider again the problem (mentioned in 1.6) of defining z^λ, where λ is real and not an integer. If w is any logarithm of z, then $|e^{\lambda w}|=|z|^\lambda$. But (firstly) there is no integer n for which $|z|^{\lambda-n}$ is bounded away from zero on some $D'(0,r)$, and (secondly) $|z|^\lambda$ does not come arbitrarily close to all real values on $D'(0,r)$. Hence there is no way of choosing logarithms so as to make $z\mapsto z^\lambda$ differentiable on a set $D'(0,r)$.

Characterization of rational functions

A **rational function** is a quotient of two polynomials. Sums, products and quotients of rational functions are again rational functions. The rational function p/q is defined and differentiable except at the zeros of q, where it has poles (not essential singularities). If the degrees of p, q are m, n, then p/q behaves like

z^{m-n} as $z \to \infty$. With the aid of the theory we have developed, we can now prove that these properties characterize rational functions:

2.5.10. *Suppose that f is defined and differentiable on $\mathbf{C}$ except at a finite number of poles, and that there exist an integer n and $k > 0$, $R > 0$ such that $|f(z)| \leqslant k|z|^n$ whenever $|z| > R$. Then f is a rational function.*

Proof. Let the poles be $z_1, \ldots, z_k$, and let their orders be $\alpha_1, \ldots, \alpha_k$ respectively. Let

$$q(z) = (z - z_1)^{\alpha_1} \cdots (z - z_k)^{\alpha_k}.$$

By removing the singularities at $z_1, \ldots, z_k$, we make fq an entire function. Applying 2.3.3 to the polynomial q, we see that there exist an integer r and $k' > 0$, $R' > 0$ such that $|f(z)q(z)| \leqslant k'|z|^r$ whenever $|z| > R'$. Hence, by 2.3.2, fq is a polynomial. The result follows.

Exercises 2.5

1 Let $f(z) = 1/z(z^2 + 1)$. Write down the Laurent series for f on $D'(0,1)$ and $\{z : |z| > 1\}$.

2 If f is an even (odd) function that does not have an essential singularity at 0, show that $\operatorname{ord}(f,0)$ is even (odd).

3 List and classify the singularities of the following functions, giving the orders of poles:

$$\frac{1}{z^2} + \frac{1}{z^2+1}, \qquad \frac{z}{\sin z}, \qquad e^{z+(1/z)}, \qquad 1/(e^{z^2} - 1).$$

4 If f has an essential singularity at z_0, show that f^2 also has. If f is non-zero in a neighbourhood of z_0, show that $1/f$ has an essential singularity at z_0.

5 Show by an example that the statement

$$a_{-n} + \cdots + a_n \to s \qquad \text{as } n \to \infty$$

is not the same as $\sum_{n=-\infty}^{\infty} a_n = s$.

6 If f is differentiable on $\mathbf{C}$ except for singularities, show that the number of singularities in any compact set is finite, and that the set of all singularities is finite or countable.

7 Let B be a bounded subset of $\mathbf{C}$. Prove that exp maps $\mathbf{C}\backslash B$ onto $\mathbf{C}\backslash\{0\}$. Deduce that cos maps $\mathbf{C}\backslash B$ onto $\mathbf{C}$.

8 Suppose that f is an entire function, and that there exist $k>0$, $R>0$ and a positive integer n such that $|f(z)| \geqslant k|z|^n$ whenever $|z|>R$. Prove that f is a polynomial. What can be said about the degree of f?

9 If f has an essential singularity at z_0, show that, for each $r>0$, there is a point z of $D'(z_0,r)$ for which $f(z)$ is real.

10 Suppose that f is differentiable on $\mathbf{C}$ except at a finite number of points, and that $zf(z)\to 0$ as $z\to\infty$. Prove that $\{z^2 f(z): |z|>R\}$ is bounded for some $R>0$.

11 Suppose that $\sum_{n=-\infty}^{\infty} a_n = A$ and $\sum_{n=-\infty}^{\infty} b_n = B$, and that $\sum_{n=-\infty}^{\infty} |a_n|$ and $\sum_{n=-\infty}^{\infty} |b_n|$ are convergent. Prove that, for each n, $\sum_{k=-\infty}^{\infty} a_k b_{n-k}$ is convergent, say to c_n, and that $\sum_{n=-\infty}^{\infty} c_n = AB$. (Let $A_n = a_{-n}+\cdots+a_n$, etc., and show that for $k, l>2n$, $c_{-k}+\cdots+c_l$ is close to $A_n B_n$.) Deduce that the Laurent series for a product of two functions is the series obtained by formally multiplying their Laurent series (and make this statement precise).

By considering the Laurent series for $e^{z+(1/z)}$ at 0, prove that

$$\frac{1}{2\pi}\int_0^{2\pi} e^{2\cos t}\cos nt\,dt = \sum_{k=0}^{\infty}\frac{1}{k!\,(k+n)!} \qquad (n\geqslant 0).$$

2.6. The residue theorem

In this section we derive our culminating theorem on complex integration. Cauchy's integral theorem and formula are easily recognizable as special cases. All the remaining material of this book, except section 3.4, consists of applications of this theorem.

Suppose that $f(z) = \sum_{n=-\infty}^{\infty} a_n(z-z_0)^n$ for $0<|z-z_0|<R$. As shown in 2.5.3, we then have $\int_{C(z_0,r)} f = 2\pi i a_{-1}$ for $0<r<R$.

Because of this, the coefficient a_{-1} is of special importance in evaluating integrals. It is called the **residue** of f at z_0, and will be denoted by $\operatorname{res}(f, z_0)$.

Winding numbers

Let φ be a closed, rectifiable path, and let z be a point not in φ^*. The **winding number** of φ with respect to z (to be denoted by $\mathrm{w}(\varphi, z)$) is defined to be

$$\frac{1}{2\pi i}\int_{\varphi}\frac{1}{\zeta - z}\,d\zeta$$

If φ is $C(a,r)$, then 2.1.5 shows that $\mathrm{w}(\varphi, z)$ is 1 if $|z-a| < r$ (i.e. if z is inside φ), and 0 if $|z-a| > r$ (i.e. if z is outside φ). By making suitable modifications to 2.1.4, one sees that similar statements hold for rectangles and semicircles. In general, we say that a closed path φ is **simple** if, for every point z not in φ^*, $\mathrm{w}(\varphi, z)$ is either 1 or 0. The points z for which $\mathrm{w}(\varphi, z) = 1$ are then said to be **inside** φ.

If φ is a closed, piecewise-smooth path in a star-shaped open set G, and $z \notin G$, then Cauchy's theorem shows at once that $\mathrm{w}(\varphi, z) = 0$.

Now let φ be any closed, piecewise-smooth path. Since φ^* is compact, it is contained in $D(0, R)$ for some $R > 0$. The preceding remark then shows that $\mathrm{w}(\varphi, z) = 0$ whenever $|z| \geqslant R$. Hence $\{z : \mathrm{w}(\varphi, z) \neq 0\}$ is bounded. In particular, the inside of a simple, closed path is always bounded.

In applications, we will always be using paths that are easily seen to be simple. Because of this, we defer a more thorough examination of winding numbers to section 3.4. There it will be shown that the winding number – as the term suggests – can be interpreted as the number of times the path 'goes round' the point (and, in particular, is always an integer). The reader who prefers to do so could read 3.4 before going on to the rest of the present section.

The residue theorem

We are now ready to state the main theorem of this section. It generalizes Cauchy's theorem to functions that are differentiable except at a finite number of points, and reduces the computation of integrals to computation of residues and winding numbers.

2.6.1 Theorem. *Let G be a star-shaped open set, and let φ be a closed, piecewise-smooth path in G. Let f be defined and differentiable on $G \backslash F$, where $F = \{z_1, \ldots, z_n\}$ is a finite set disjoint from φ^*. Then*

$$\int_\varphi f = 2\pi i \sum_{j=1}^{n} \operatorname{res}(f, z_j) \operatorname{w}(\varphi, z_j).$$

Proof. Let f_j be the principal part of f at z_j, and let $g = f - (f_1 + \cdots + f_n)$. Now $f_2 + \cdots + f_n$ is differentiable at z_1, and $f - f_1$ has a removable singularity there. Hence g has a removable singularity at z_1, and similarly at each z_j. With these singularities removed, g becomes differentiable on G. Therefore, by Cauchy's theorem, $\int_\varphi g = 0$, or

$$\int_\varphi f = \sum_{j=1}^{n} \int_\varphi f_j.$$

Since φ^* is a closed set, there exists, for each j, $\delta_j > 0$ such that $|z - z_j| \geqslant \delta_j$ for z in φ^*. Let the Laurent series for f at z_j be $\sum_{n=-\infty}^{\infty} a_n(z - z_j)^n$. Then $f_j(z) = \sum_{n=1}^{\infty} a_{-n}(z - z_j)^{-n}$ for z in $\mathbf{C} \backslash \{z_j\}$, the series being uniformly convergent on $\{z : |z - z_j| \geqslant \delta_j\}$. Hence

$$\int_\varphi f_j = \sum_{n=1}^{\infty} a_{-n} \int_\varphi (z - z_j)^{-n} \, dz = a_{-1} \, 2\pi i \operatorname{w}(\varphi, z_j).$$

The result follows.

In particular, if φ is a simple closed path, then

$$\int_\varphi f = 2\pi i \times (\text{sum of residues inside } \varphi),$$

a form in which the statement of the residue theorem is especially memorable.

Evaluation of residues at poles

It is now evident that it is important to know how to compute residues. The rest of this section contains some rules for their computation at poles. Although these results are rather special they are still very useful in applications. The first result is almost obvious.

2.6.2. *If* $\operatorname{ord}(f, z_0) = -1$, *then* $\operatorname{res}(f, z_0) = \lim_{z \to z_0} (z - z_0) f(z)$.

Proof. Let the Laurent series for f at z_0 be

$$f(z) = \frac{a_{-1}}{z - z_0} + \sum_{n=0}^{\infty} a_n (z - z_0)^n.$$

Then

$$\begin{aligned}(z - z_0) f(z) &= a_{-1} + \sum_{n=0}^{\infty} a_n (z - z_0)^{n+1} \\ &\to a_{-1} \qquad \text{as } z \to z_0.\end{aligned}$$

2.6.3 Corollary. *If* $\operatorname{ord}(f, z_0) = -1$ *and* $\operatorname{ord}(g, z_0) = 0$, *then* $\operatorname{res}(fg, z_0) = g(z_0) \operatorname{res}(f, z_0)$.

Proof. $\operatorname{ord}(fg, z_0) = -1$, by 2.5.5. Hence

$$\operatorname{res}(fg, z_0) = \lim_{z \to z_0} (z - z_0) f(z) g(z) = g(z_0) \operatorname{res}(f, z_0).$$

2.6.4 Corollary. *If* $\operatorname{ord}(f, z_0) = 0$ *and* $\operatorname{ord}(g, z_0) = 1$, *then*

$$\operatorname{res}\left(\frac{f}{g}, z_0\right) = \frac{f(z_0)}{g'(z_0)}.$$

Proof. $\operatorname{ord}(f/g, z_0) = -1$, by 2.5.5, so

$$\operatorname{res}\left(\frac{f}{g}, z_0\right) = \lim_{z \to z_0} (z - z_0) \frac{f(z)}{g(z)} = \frac{f(z_0)}{g'(z_0)}.$$

Notice that 2.6.4 is applicable whenever $g(z_0) = 0$ and both parts of the fraction are non-zero. For $f(z_0) \neq 0$ implies that $\operatorname{ord}(f, z_0) = 0$, and $g'(z_0) \neq 0$ that $\operatorname{ord}(g, z_0) = 1$.

Examples. (i) The residue of $z \mapsto 1/(1 + z^2)$ at i is $1/2i$.

(ii) The function $z \mapsto 1/\sin z$ has a simple pole at $n\pi$ for each integer n. The residue at $n\pi$ is $1/\cos n\pi = (-1)^n$. Hence the integral of $1/\sin z$ round the unit circle is $2\pi i$.

(iii) It is always easy to apply 2.6.2 to a simple pole of a ratio of two series. To illustrate this, let

$$f(z) = \frac{\sin z}{1 - \cos z} = \left(1 - \frac{z^2}{3!} + \cdots\right) \Big/ \left(\frac{z}{2!} - \frac{z^3}{4!} + \cdots\right)$$

Then f has a simple pole at 0, and, by 1.2.7,

$$\operatorname{res}(f, 0) = \lim_{z \to 0} z f(z) = 2.$$

The next result is a generalized form of Cauchy's integral formula, in which the integral is taken round an arbitrary closed, piecewise-smooth path instead of a circle.

2.6.5. *Suppose that f is differentiable on a star-shaped open set G, and that φ is a closed, piecewise-smooth path in G. Then, for z in $G \backslash \varphi^*$,*

$$\int_\varphi \frac{f(\zeta)}{\zeta - z} d\zeta = 2\pi i f(z) \, \mathrm{w}(\varphi, z).$$

Proof. Let $F(\zeta) = f(\zeta)/(\zeta - z)$ $(\zeta \in G \backslash \{z\})$. By 2.6.2, $\operatorname{res}(F, z) = f(z)$. The result follows, by the residue theorem.

Returning to the problem of evaluating residues, we can give the following rule for poles of higher orders:

2.6.6. *If* $\operatorname{ord}(f, z_0) = -k$ *(where* $k > 0$*), then the residue of f at z_0 is* $h^{(k-1)}(z_0)/(k-1)!$, *where* $h(z) = (z - z_0)^k f(z)$ *(and the singularity of h at z_0 is removed).*

Proof. Let the Laurent series for f at z_0 be $\sum_{n=-k}^{\infty} a_n (z - z_0)^n$, so that $h(z) = a_{-k} + \cdots + a_{-1}(z - z_0)^{k-1} + \cdots$. Then $h^{(k-1)}(z_0) = (k-1)!\, a_{-1}$.

Examples. (i) Let $f(z) = z^2/(z^2 + a^2)^2$, where $a \neq 0$. Then f has a double pole at ia. Let

$$h(z) = (z - ia)^2 f(z) = z^2/(z + ia)^2.$$

Then $h'(z) = 2iaz/(z + ia)^3$, so $\operatorname{res}(f, ia) = h'(ia) = 1/4ia$.

(ii) Sometimes use of 2.6.6 will involve very heavy computation, which can be avoided with a little ingenuity. For instance, if f has only one non-zero Laurent coefficient before a_{-1}, it may be easier to calculate this coefficient and a_{-1} by taking limits. To illustrate this, let $f(z) = 1/(z^2 \sin z)$. Then $\operatorname{ord}(f,0) = -3$, and $a_{-3} = \lim_{z\to 0} (z/\sin z) = 1$. Now

$$f(z) - \frac{1}{z^3} = \frac{z - \sin z}{z^3 \sin z} = \frac{1}{\sin z}\left(\frac{1}{3!} - \frac{z^2}{5!} + \cdots\right).$$

Hence $z(f(z) - 1/z^3) \to \frac{1}{6}$ as $z \to 0$, showing that $a_{-2} = 0$ and $a_{-1} = \frac{1}{6}$.

The occasional short cut should not be missed; for example, the residue at 0 of any even function is 0.

Exercises 2.6

1 Prove that

$$\int_{C(0,\,2)} \frac{e^{az}}{1+z^2}\,dz = 2\pi i \sin a.$$

2 If $\operatorname{ord}(f, z_0) = k > 0$ and $\operatorname{ord}(g, z_0) = k+1$, show that

$$\operatorname{res}\left(\frac{f}{g}, z_0\right) = (k+1)\frac{f^{(k)}(z_0)}{g^{(k+1)}(z_0)}.$$

3 Find the residues of the following functions at the points stated:

(i) $\dfrac{e^{iz}}{z^3+z}$ at i, 1; (ii) $\dfrac{1}{1-\cos z}$ at 0;

(iii) $\dfrac{1}{(z-1)^2(z+1)}$ at 1 (ans. $\frac{1}{4}$);

(iv) $\dfrac{1}{z-\sin z}$ at 0 (ans. $\frac{3}{10}$).

4 Show by an example that the conclusion of 2.6.3 does not hold if f has a pole of higher order.

5 Suppose that f is differentiable on $\mathbf{C}\backslash\{z_1,\ldots,z_n\}$, and has Laurent series $\sum_{-\infty}^{\infty} a_n z^n$ on $\{z:|z|>R\}$, where $R=\max|z_j|$. Prove that $a_{-1}=\sum_{j=1}^{n}\operatorname{res}(f,z_j)$.

2.7. Integration of f'/f and the local mapping theorem

Sections 2.2, 2.3 and 2.4 consisted of a series of results on complex functions that were all deduced from Cauchy's integral formula. We are now equipped with a stronger theorem on integration – the residue theorem – and in the present section we will see what further properties of complex functions can be deduced from it. By considering the integral of the quotient f'/f, we obtain some useful theorems concerning the location of zeros and poles, and from these we deduce a very strong result on the nature of the local mapping given by a differentiable function.

In other words, this section consists of theoretical applications of the residue theorem. Some of its applications to particular problems (often not even involving complex numbers) are considered in 3.1, 3.2 and 3.3.

2.7.1 Theorem. *Let φ be a closed, piecewise-smooth path in a star-shaped open set G. Let f be a function that is differentiable on G except at a finite number of poles, and such that $\{z\in G:\operatorname{ord}(f,z)\neq 0\}$ is a finite set $\{z_1,\ldots,z_n\}$ disjoint from φ^*. Then*

$$\frac{1}{2\pi i}\int_{\varphi}\frac{f'}{f}=\sum_{j=1}^{n}\operatorname{ord}(f,z_j)\,\mathrm{w}(\varphi,z_j).$$

Proof. The function f'/f is differentiable on $G\backslash\{z_1,\ldots,z_n\}$. Let $\operatorname{ord}(f,z_j)=p_j$. We show that $\operatorname{res}(f'/f,z_j)=p_j$. The residue theorem then gives the result.

There is a function g, differentiable and non-zero at z_j, such that $f(z)=(z-z_j)^{p_j}g(z)$ for $z\neq z_j$. Then $f'(z)=p_j(z-z_j)^{p_j-1}g(z)+(z-z_j)^{p_j}g'(z)$, so

$$\frac{f'(z)}{f(z)}=\frac{p_j}{z-z_j}+\frac{g'(z)}{g(z)}.$$

The required equality follows.

Under the conditions of the theorem, let us denote

$$\sum_{j=1}^{n} \operatorname{ord}(f, z_j)\, \mathrm{w}(\varphi, z_j)$$

by $\mathrm{ZP}(f, \varphi)$. In the case when it is a simple path, this quantity reduces to $\sum \operatorname{ord}(f, z_j)$, where only the z_j inside φ are included. It may be described as the number of zeros inside minus the number of poles, where each is counted according to its order.

Rouché's theorem

The next theorem says, roughly speaking, that ZP is unchanged by sufficiently small deformations, but the deformations allowed are, in fact, fairly large: we can add to f any function g such that $|g(z)| < |f(z)|$ for z in φ^*. This theorem sounds fairly innocuous; its power will be better appreciated when we have seen some of its applications. We give two proofs. The first one includes a reminder of an interpretation of $\int_\varphi (f'/f)$ that we have neglected hitherto: by 1.7.7, it is equal to $\int_{f\circ\varphi} (1/z)dz$ – in other words, to $2\pi i \mathrm{w}(f \circ \varphi, 0)$.

2.7.2 Theorem. *Let φ be a closed, piecewise-smooth path in a star-shaped open set G. Suppose that f and g are functions on G such that:*

(i) *f and g are differentiable on G, except at a finite number of poles, none of which lie on φ^*;*

(ii) *f and $f+g$ have at most a finite number of zeros in G;*

(iii) *for z in φ^*, $|g(z)| < |f(z)|$.*

Then $\mathrm{ZP}(f+g, \varphi) = \mathrm{ZP}(f, \varphi)$.

Proof 1. Let

$$h(z) = \frac{f(z)+g(z)}{f(z)} = 1 + \frac{g(z)}{f(z)}.$$

By (iii), f and $f+g$ are non-zero on φ^*, so h is differentiable and non-zero on φ^*. By 2.5.5,

$$\operatorname{ord}(h, z) = \operatorname{ord}(f+g, z) - \operatorname{ord}(f, z) \qquad (z \in G),$$

so, summing over the points where f or $f+g$ has non-zero order, $\mathrm{ZP}(h, \varphi) = \mathrm{ZP}(f+g, \varphi) - \mathrm{ZP}(f, \varphi)$. By 2.7.1, $\mathrm{ZP}(h, \varphi) = \int_\varphi (h'/h)$, and by 1.7.7, this is equal to $\int_{h\circ\varphi}(1/z)dz$. But $|h(z)-1)<1$ for z in φ^*, and hence $h\circ\varphi$ is a closed path in the convex set $D(1,1)$, on which $z\mapsto 1/z$ is differentiable. Therefore $\int_{h\circ\varphi}(1/z)dz=0$, by Cauchy's theorem. The result follows.

Proof 2. (This proof uses the fact (3.4.2) that winding numbers are integers. For the moment, let us regard it as applying only to simple paths.) Since φ^* is compact,

$$\inf\{|f(z)|-|g(z)|: z\in\varphi^*\}=\delta>0.$$

For $0\leqslant t\leqslant 1$, define

$$E(t)=\frac{1}{2\pi i}\int_{\varphi}\frac{f'+tg'}{f+tg}.$$

This is defined because $f+tg$ is non-zero on φ^*, and by 2.7.1, it is an integer. If $0\leqslant t<u\leqslant 1$ and $z\in\varphi^*$, then

$$\left|\frac{(f'+ug')(z)}{(f+ug)(z)}-\frac{(f'+tg')(z)}{(f+tg)(z)}\right|\leqslant\frac{u-t}{\delta^2}|(f'g-fg')(z)|.$$

Hence $|E(u)-E(t)|\leqslant k(u-t)$, where k is a constant independent of t and u. If $u-t<1/k$, then $|E(u)-E(t)|<1$, so, since $E(u)$ and $E(t)$ are integers, $E(u)=E(t)$. Choosing an integer $n>k$, it follows that $E(r/n)=E(0)$ for $r=1,\ldots,n$. In particular, $E(1)=E(0)$, which proves the theorem.

Rouché's theorem gives an alternative proof that a polynomial of degree n has n zeros, if counted according to order. For if $p(z)=q(z)+a_n z^n$, then $|q(z)|<|a_n z^n|$ for sufficiently large $|z|$, and $\mathrm{ZP}(a_n z^n, C)=n$ for any circle C with centre 0.

Example. As another application of Rouché's theorem, we prove that if $|a|>e$ and $n\geqslant 1$, then there are n distinct values of z such that $\operatorname{Re} z<\log|a|$ and $e^z=az^n$, and that each lies inside the unit circle.

Take $R>1$ and μ such that $1\leqslant\mu<\log|a|$. Let φ be the rectangle with vertices $\mu\pm Ri$, $-R\pm Ri$. Now $|e^z|<|a|$ whenever $\operatorname{Re} z<\log|a|$, from which we see that $|az^n|>|e^z|$ both for $|z|=1$ and for z in φ^*. Hence az^n-e^z has the same number of zeros (counted with orders) as az^n inside both $C(0,1)$ and φ, that is, n

in each case. Given any z with $\operatorname{Re} z < \log|a|$, a suitable choice of R and μ ensures that z is inside φ, so these n zeros are the only ones with $\operatorname{Re} z < \log|a|$. The zeros are all simple, and therefore distinct, for $az^n - e^z$ and its derivative $naz^{n-1} - e^z$ do not vanish together.

The local mapping theorem

The function $z \mapsto z^n$ maps each neighbourhood of 0 on to another neighbourhood of 0 in an 'n-to-one' fashion, in the following sense: if $\epsilon > 0$, and $0 < |a| < \epsilon^n$, then there are n distinct points z in $D(0, \epsilon)$ such that $z^n = a$. Our next theorem asserts that every non-constant, differentiable complex function has a similar property. More precisely, suppose that $f(z_0) = a_0$. Then $f - a_0$ has a zero at a_0: let n be the order of this zero. We prove that each neighbourhood of z_0 is mapped in an n-to-one fashion on to some neighbourhood of a_0. The exact statement is as follows:

2.7.3 Theorem. *Suppose that f is non-constant and differentiable on a neighbourhood of z_0, and write $f(z_0) = a_0$. Let n be the order of the zero of $f - a_0$ at z_0. If $\epsilon > 0$ is sufficiently small, then there is a corresponding $\delta > 0$ such that for each a in $D'(a_0, \delta)$, there exist n distinct points $z_1, \ldots, z_n$, of $D'(z_0, \epsilon)$ with $f(z_j) = a$. The zero of $f - a$ at each z_j is simple.*

In particular, f maps each neighbourhood of z_0 on to a neighbourhood of a_0.

Proof. By 2.2.8, if $\epsilon > 0$ is sufficiently small, then for $0 < |z - z_0| < 2\epsilon$, $f(z)$ is defined, and (i) $f(z) \neq a_0$, (ii) $f'(z) \neq 0$. Since $\{z: |z - z_0| = \epsilon\}$ is compact,

$$\inf\{|f(z) - a_0| : |z - z_0| = \epsilon\} = \delta > 0.$$

Choose a in $D'(a_0, \delta)$. Now

$$f(z) - a = (f(z) - a_0) + (a_0 - a),$$

and we have arranged that $|a_0 - a| < |f(z) - a_0|$ whenever $|z - z_0| = \epsilon$. Rouché's theorem therefore shows that $f - a$ has n zeros (counted with orders) in $D(z_0, \epsilon)$. Since f' is non-zero on this disc, these n zeros are all simple, and therefore distinct.

Any neighbourhood N of z_0 contains $D(z_0, \epsilon)$ for some ϵ satisfying the above conditions. Then, with the above notation, $f(N)$ contains $D(a_0, \delta)$.

This information about the nature of a differentiable complex function puts various earlier results into perspective. The maximum modulus theorem (2.4.1) follows trivially, since every neighbourhood of a_0 contains points with modulus greater than $|a_0|$. The fact (1.5.12) that a non-constant, differentiable complex function cannot map an open set into $\mathbf{R}$ is another obvious consequence, and it is now clear that the same is true if $\mathbf{R}$ is replaced by any set having empty interior in $\mathbf{C}$. Another immediate deduction is exercise 9 of 2.2.

If f is one-to-one on G, with inverse g, then 2.7.3 says the following: given $\epsilon > 0$, there exists $\delta > 0$ such that for all a in $D(a_0, \delta)$, $|g(a) - g(a_0)| < \epsilon$. In other words, g is continuous. Furthermore, 2.7.3 tells us under what conditions f is one-to-one on some neighbourhood of z_0: this occurs if and only if $\text{ord}(f - a_0, z_0) = 1$, i.e. if and only if $f'(z_0) \neq 0$. We summarize these conclusions in the following 'inverse function theorem':

2.7.4 Corollary. *Suppose that f is differentiable on a neighbourhood of z_0, and that $f'(z_0) \neq 0$. Then there exists $\epsilon' > 0$ such that f is one-to-one on N, where $N = D(z_0, \epsilon')$. If g is the inverse function to $f|_N$, $z \in N$ and $f(z) = a$, then $g'(a) = 1/f'(z)$.*

Proof. Write $f(z_0) = a_0$, and let ϵ, δ be as in 2.7.3. Since f is continuous at z_0, there exists ϵ' such that $0 < \epsilon' \leqslant \epsilon$ and $|f(z) - a_0| < \delta$ whenever $|z - z_0| < \epsilon'$. Let $N = D(z_0, \epsilon')$. Take z_1, z_2 in N, and suppose that $f(z_1) = f(z_2) = a$. By 2.7.3, only one point in N is mapped to a, and so $z_1 = z_2$. Hence f is one-to-one on N. The inverse g of $f|_N$ is continuous, by the remark above, and the last statement now follows from 1.5.5.

We only assert, in 2.7.4, that f is one-to-one on *some neighbourhood* of z_0. A function may have a non-zero derivative on a connected open set G without being one-to-one on G. For example, the exponential function has a non-zero derivative on all of $\mathbf{C}$, but, as we know, is not one-to-one. In fact (paradoxical as this may seem after 2.7.4), the local mapping theorem enables us to prove that there are very few one-to-one entire functions:

2.7.5. *If f is a one-to-one entire function, then there exist a, b in* $\mathbf{C}$ $(a \neq 0)$ *such that* $f(z) = az + b$ $(z \in \mathbf{C})$.

Proof. If f is not a polynomial, then $\{f(z):|z| > 1\}$ is dense in C, by 2.5.9. But $\{f(z):|z| < 1\}$ is a neighbourhood of $f(0)$, by 2.7.3, and therefore meets $\{f(z):|z| > 1\}$. This contradicts the hypothesis that f is one-to-one, and hence f is a polynomial. Since f has only one zero, it is of the form $a(z - z_0)^n$ for some a, b, z_0. If $n > 1$, we then have $f(z_0 + \omega) = f(z_0 + \omega')$, where ω and ω' are two distinct nth roots of one. Hence $n = 1$, giving the result.

For real functions, 2.7.4 is almost trivial (if continuity of the derivative is assumed). For if $f'(x_0) > 0$, then $f'(x) > 0$ for x in a neighbourhood of x_0, and the mean-value theorem shows that f is strictly increasing in this neighbourhood. Similar reasoning applies if $f'(x_0) < 0$. On the other hand, 2.7.3 is quite false for real functions, as is shown by the example of $x \mapsto x^2$ at 0: the function does not take values less than 0.

A similar theorem to 2.7.4 holds for functions that are differentiable as functions on $\mathbf{R}^2$ (or $\mathbf{R}^n$), see e.g. Spivak [7], pages 35–8.

Hurwitz's theorem

Another application of Rouché's theorem is the following result on sequences of functions:

2.7.6 (Hurwitz). *Suppose that $\{f_n\}$ is a sequence of functions, each differentiable on an open set G, and that $\{f_n\}$ converges to a non-constant function f, uniformly on compact subsets of G. Choose z_0 in G, and write $f(z_0) = a_0$. Then, given $\epsilon > 0$, there is an integer $N(\epsilon)$ such that for each $n \geqslant N(\epsilon)$, there exists z_n with $|z_n - z_0| < \epsilon$ and $f_n(z_n) = a_0$.*

Proof. We know from 2.2.10 that f is differentiable. By 2.2.6, there exists $r > 0$ such that $f(z) \neq a_0$ whenever $0 < |z - z_0| < r$. It is sufficient to prove the theorem for $\epsilon < r$. Choosing such an ϵ, we have

$$\inf\{|f(z)-a_0|:|z-z_0|=\epsilon\}=\delta>0.$$

There exists $N(\epsilon)$ such that whenever $n \geqslant N(\epsilon)$, $|f_n(z)-f(z)|<\delta$ for all z with $|z-z_0|=\epsilon$. For $n \geqslant N(\epsilon)$ and $|z-z_0|=\epsilon$, we have $|f(z)-a_0|>|f_n(z)-f(z)|$. Since $f-a_0$ has a zero in $D(z_0,\epsilon)$, the same is true of f_n-a_0, by Rouché's theorem.

One consequence of this theorem is that each value of f is a value of some f_n. Also, we have:

2.7.7 Corollary. *Suppose that $\{f_n\}$ is a sequence of functions, each differentiable on an open set G, and that $\{f_n\}$ converges to a non-constant function f, uniformly on compact subsets of G. If each f_n is one-to-one, then so is f.*

Proof. Suppose that $z_1 \neq z_2$ and that $f(z_1)=f(z_2)=a$. Take $\epsilon<\frac{1}{2}|z_1-z_2|$. Then there exist n and z_1', z_2' such that $|z_j'-z_j|<\epsilon$ and $f_n(z_j')=a_j$ $(j=1,2)$. Since $z_1' \neq z_2'$, this shows that f_n is not one-to-one.

Exercises 2.7

1 Under the conditions of 2.7.1, prove that if g is differentiable on G, then

$$\int_\varphi \frac{f'g}{f}=2\pi i\sum_{j=1}^{n} g(z_j)\operatorname{ord}(f,z_j)\operatorname{w}(\varphi,z_j).$$

2 If f is differentiable and non-constant on an open set G, and H is an open subset of G, show that $f(H)$ is open.

3 Show by an example that Hurwitz's theorem is false for real functions.

4 Prove that there are exactly three distinct values of z such that $2\frac{1}{2}<|z|<3$ and $z^4+26z+2=0$.

5 By considering the moduli of $\cos z$ and $1/z$ on the sides of the rectangles with vertices $\pm\pi\pm iR$ for all sufficiently large R, find how many zeros of $\cos z - 1/z$ (counted with their orders) lie in $\{z: -\pi < \operatorname{Re} z < \pi\}$. How many of these zeros are real? By considering conjugates (or otherwise), prove that the non-real zeros are all simple, and therefore distinct.

6 Using either proof, show that Rouché's theorem still holds if condition (iii) is replaced by the following: $f(z) + tg(z) \neq 0$ whenever $z \in \varphi^*$ and $0 \leqslant t \leqslant 1$.

If m, n are positive integers, m being odd and n even, prove that there are $m + n$ distinct values of z such that $-m\pi < \operatorname{Re} z < n\pi$ and $\cos z = z$.

7 Let f be defined and differentiable on **C**, except at isolated singularities, and suppose that f is one-to-one. Prove that:

(i) f has no essential singularities;

(ii) f has at most one pole (Hint: $1/f$ is one-to-one);

(iii) the Laurent series for f at its pole (if it has one) has only a finite number of non-zero coefficients;

(iv) if f is not entire, then it is of the form

$$f(z) = \frac{az + b}{z - z_0}.$$

8 Make a list of the results in chapter 2 that have been proved using integration, but do not mention integration in their statement.

CHAPTER THREE

Further topics

The four sections of this chapter are logically independent, and can be read in any order. The first three are concerned with computations that can be performed with the aid of the residue theorem, and contain an impressive body of applications of complex function theory to problems on real numbers (cf. the remarks in 2.3). The fourth section is devoted to winding numbers, supplementing the very brief treatment in 2.6.

3.1. The evaluation of real integrals

The residue theorem provides a very effective method of evaluating certain real integrals. The general idea is to choose a closed path in $\mathbf{C}$ and a complex function so that the integral over one part of the path is the required real integral, while the contributions of the other parts are either known or small. In using the residue theorem to write down the value of the integral round the whole path, we shall make repeated use of the rules for computation of residues given in 2.6, and we shall omit the formal verification of the required facts about winding numbers. For the paths considered (semicircles, rectangles, etc.), this will always be possible by making slight modifications to the proof of 2.1.4.

Most of the real integrals for which the method works are integrals over an unbounded range, and we start with the precise definition of such integrals. Let f be a real function such that for each $x > a$, $\int_a^x f$ exists. Write $F(x) = \int_a^x f$. If $F(x)$ tends to a finite limit L as $x \to \infty$, then we write $\int_a^\infty f = L$, and say that the integral $\int_a^\infty f$ 'converges'. Similarly, we define $\int_{-\infty}^a f$ to be $\lim_{x \to -\infty} \int_x^a f$ if this exists. Finally, if both $\int_{-\infty}^0 f$ and $\int_0^\infty f$ exist, we write

$$\int_{-\infty}^{\infty} f = \int_{-\infty}^{0} f + \int_{0}^{\infty} f.$$

Example. Let $f(x) = 1/x^2$ $(x \neq 0)$. Then $\int_1^x f = 1 - 1/x$, so $\int_1^\infty f = 1$.

In dealing with convergence of integrals, one often needs the following elementary consequence of the completeness of $\mathbf{R}$: If F is a real function, and for each $\epsilon > 0$, there exists R such that $|F(x) - F(y)| \leqslant \epsilon$ whenever $x, y > R$, then $F(x)$ tends to a limit as $x \to \infty$.

The proof of the following statements is almost immediate:

3.1.1. (i) *If f is continuous and $\int_a^\infty |f|$ exists, then $\int_a^\infty f$ exists.*

(ii) *If f and g are continuous, $\int_a^\infty g$ exists, and there is a positive number k such that $0 \leqslant f(x) \leqslant kg(x)$ for $x \geqslant a$, then $\int_a^\infty f$ exists.*

3.1.2. *The following statement is equivalent to $\int_{-\infty}^\infty f = s$: given $\epsilon > 0$, there exists $R > 0$ such that, whenever $u, v \geqslant R$, $|\int_u^v f - s| \leqslant \epsilon$.*

Proof. (i) Suppose that $\int_{-\infty}^\infty f = s$, so that $s = s_1 + s_2$, where $\int_{-\infty}^0 f = s_1$ and $\int_0^\infty f = s_2$. There exists $R > 0$ such that, whenever $u, v \geqslant R$, $|\int_{-u}^0 f - s_1| \leqslant \epsilon/2$ and $|\int_0^v f - s_2| \leqslant \epsilon/2$. For such u, v, we have $|\int_{-u}^v f - s| \leqslant \epsilon$.

(ii) Suppose that the condition holds. Take $\epsilon > 0$, and let R correspond to ϵ as stated. If $x, y > R$, then

$$\left|\int_x^y f\right| = \left|\int_{-R}^y f - \int_{-R}^x f\right| \leqslant 2\epsilon.$$

Hence $\int_0^\infty f$ converges. Similarly, $\int_{-\infty}^0 f$ converges. Let $\int_{-\infty}^\infty f = s'$. By (i), there exists $u \geqslant R$ such that $|\int_{-u}^u f - s'| \leqslant \epsilon$. Hence $|s' - s| \leqslant 2\epsilon$. This is true for all $\epsilon > 0$, so $s' = s$.

Note. If $\int_{-x}^{x} f$ tends to a limit as $x \to \infty$, it does not follow that $\int_{-\infty}^{\infty} f$ exists, a point that is often overlooked. An obvious counter-example is provided by the function $f(x) = x$. However, if we know that $\int_{-\infty}^{\infty} f$ exists, then 3.1.2 shows that it is equal to $\lim_{x\to\infty} \int_{-x}^{x} f$.

The basic method

The most straightforward applications of the residue theorem to the evaluation of real integrals are those in which the closed path consists of part of the real axis (giving an approximation to the required integral) and a return path on which the integral tends to zero. We illustrate this by evaluating $\int_{-\infty}^{\infty} [x^2/(x^2+a^2)^2]\,dx$, where $a > 0$. Convergence of the integral is immediate on comparison with $1/x^2$. Let $f(z) = z^2/(z^2+a^2)^2$. For $r > 0$, let $\varphi(r)$

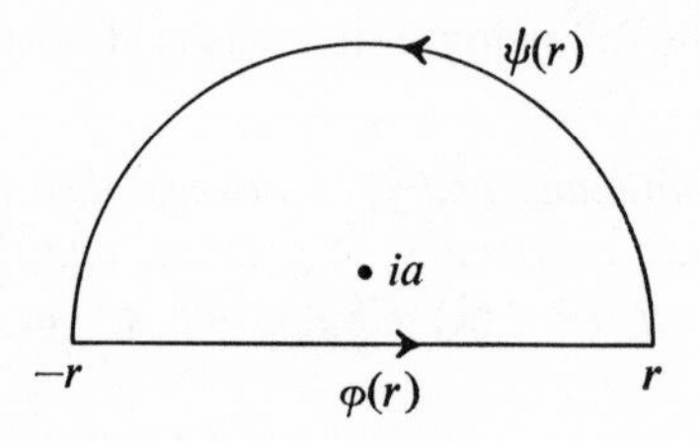

be the real segment $[-r \to r]$, and let $\psi(r)$ be the semicircle $t \mapsto re^{it}$ $(0 \leqslant t \leqslant \pi)$. Then

$$\int_{\varphi(r)} f = \int_{-r}^{r} f,$$

while

$$\left|\int_{\psi(r)} f\right| \leqslant \frac{\pi r^3}{(r^2-a^2)^2}$$

$$\to 0 \qquad \text{as } r \to \infty$$

Hence $\int_{\varphi(r)\cup\psi(r)} f \to \int_{-\infty}^{\infty} f$ as $r \to \infty$. Now f has poles at $\pm ia$. The winding number is 1 at ia and 0 at $-ia$. By example (i) following 2.6.6, $\operatorname{res}(f, ia) = 1/4ia$, so $\int_{\varphi(r)\cup\psi(r)} f = \pi/2a$ for all r. Hence $\int_{-\infty}^{\infty} f = \pi/2a$.

The reader will soon appreciate the power of this method if he attempts to evaluate the above integral by writing down an antiderivative. In fact, it is known that some of the functions considered in this section do not possess antiderivatives that can be expressed as combinations of the so-called 'elementary' functions.

We now state as generally as possible the result that is established by the method of the above example:

3.1.3. *Suppose that f is differentiable on* **C** *except at a finite number of points, none being on the real line, and those in $\{z : \operatorname{Im} z > 0\}$ being $a_1, \ldots, a_n$. Suppose, also, that positive numbers M, R exist such that $|z^2 f(z)| \leqslant M$ whenever $\operatorname{Im} z \geqslant 0$ and $|z| > R$. Then*

$$\int_{-\infty}^{\infty} f = 2\pi i \sum_{j=1}^{n} \operatorname{res}(f, a_j).$$

Proof. We need only point out that the condition on $|z^2 f(z)|$ ensures that $\int_{-\infty}^{\infty} f$ converges, and that the integral round the semicircle of radius r tends to zero as $r \to \infty$.

Notice that this result can be applied to all rational functions of the form p/q, where q has no real zeros and the degree of q exceeds the degree of p by at least two (the required inequality then holds, by 2.3.3). The degree of q must, of course, be even for the first of these conditions to be satisfied.

The method also applies to products of rational and trigonometric functions. For $a > 0$, we have

$$|e^{ia(x+iy)}| = e^{-ay} \leqslant 1 \qquad (y \geqslant 0),$$

so if f satisfies the conditions of 3.1.3, then $e^{iaz} f(z)$ also does so. Provided that f maps **R** into **R**, consideration of the real and imaginary parts therefore gives the values of

$$\int_{-\infty}^{\infty} f(x) \cos ax \, dx \qquad \text{and} \qquad \int_{-\infty}^{\infty} f(x) \sin ax \, dx.$$

Notice that it would not do to integrate $f(z) \cos az$ round the semicircle, because $|\cos az|$ becomes large when $\operatorname{Im} z$ is large.

Example. To evaluate $I = \int_{-\infty}^{\infty} [\cos x/(1+x^2)] \, dx$. Let $f(z) = e^{iz}/(1+z^2)$. Then f has a simple pole at i, with residue $1/2ie$. Hence, by 3.1.3, $\int_{-\infty}^{\infty} [e^{ix}/(1+x^2)]\,dx = \pi/e$. Taking the real part, we have $I = \pi/e$.

We now show that, for products of this kind, it is sufficient to have the weaker condition that $f(z) \to 0$ as $z \to \infty$ with $\operatorname{Im} z \geqslant 0$ (by this we mean, of course, that for every $\epsilon > 0$, there exists R such that $|f(z)| \leqslant \epsilon$ whenever $|z| > R$ and $\operatorname{Im} z \geqslant 0$). Convergence of the real integral is no longer obvious in this case, and will be proved using the criterion of 3.1.2.

3.1.4 (Jordan's lemma). *Suppose that f is differentiable on $\mathbf{C}$ except at a finite number of points, none being on the real line, and those in $\{z:\operatorname{Im} z > 0\}$ being $c_1, \ldots, c_n$. Suppose, also, that $f(z) \to 0$ as $z \to \infty$ with $\operatorname{Im} z \geqslant 0$, and let a be a positive real number. Then*

$$\int_{-\infty}^{\infty} f(x)\cos ax\,dx + i\int_{-\infty}^{\infty} f(x)\sin ax\,dx = 2\pi i \sum_{j=1}^{n} \operatorname{res}(g, c_j)$$

where $g(z) = f(z)e^{iaz}$.

Proof. Given $\epsilon > 0$, take R such that (i) $|c_j| < R$ for each j, (ii) $|f(z)| \leqslant \epsilon$ whenever $\operatorname{Im} z \geqslant 0$ and $|z| > R$, and (iii) $te^{-at} \leqslant 1$

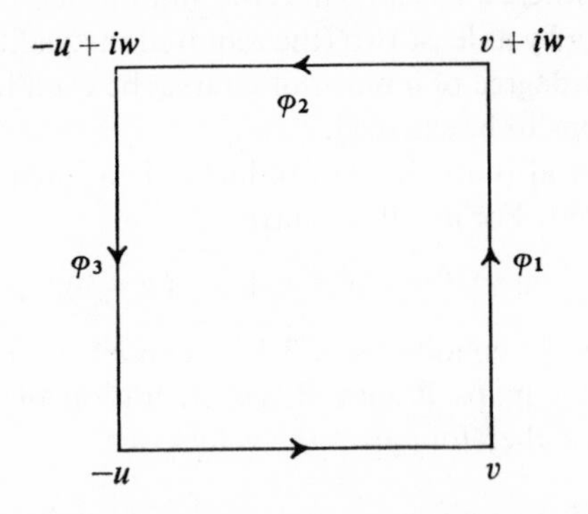

whenever $t \geqslant R$. Choose u, $v > R$, and write $w = u + v$. Let φ be the square with vertices $-u$, v, $v + iw$, $-u + iw$. Then

$$\int_{\varphi} g = 2\pi i \sum_{j=1}^{n} \operatorname{res}(g, c_j).$$

Let $\varphi_1 = [v \to v + iw]$, $\varphi_2 = [v + iw \to -u + iw]$, $\varphi_3 = [-u + iw \to -u]$. For $z = x + iy \in \varphi_j^*$ $(j = 1, 2, 3)$, we have $|f(z)| \leqslant \epsilon$ and $|e^{iaz}| = e^{-ay}$. Hence, for $j = 1$, 3,

$$\left|\int_{\varphi_j} g\right| \leqslant \epsilon \int_0^w e^{-ay}\,dy = \frac{\epsilon}{a}(1 - e^{-aw}) \leqslant \frac{\epsilon}{a}.$$

Also, since $L(\varphi_2) = w$, we have

$$\left|\int_{\varphi_2} g\right| \leqslant w\epsilon\, e^{-aw} \leqslant \epsilon.$$

The result follows, by 3.1.2.

If, in the proof of 3.1.4, we integrated round a semicircle with centre 0 (instead of a square), we would only be able to deduce that $\lim_{u\to\infty} \int_{-u}^{u} g$ exists (cf. the remark after 3.1.2).

Example. To evaluate $I = \int_{-\infty}^{\infty} [x \sin x/(x^2 + a^2)]\,dx$, where $a > 0$. Let $f(z) = z e^{iz}/(z^2 + a^2)$. The only pole of f in the upper half-plane is ia. By 2.6.4, the residue there is

$$\frac{ia\,e^{-a}}{2ia} = \tfrac{1}{2} e^{-a}.$$

By Jordan's lemma, it follows that $\int_{-\infty}^{\infty} f = \pi i e^{-a}$. Taking the imaginary part, $I = \pi e^{-a}$.

Note. Since the integrand is even, we can deduce that

$$\int_0^\infty \frac{x \sin x}{x^2 + a^2}\,dx = \tfrac{1}{2} I = \frac{\pi}{2} e^{-a}.$$

By passing from $\sin x$ or $\cos x$ to e^{iz}, we may introduce a simple pole on the path of integration. This can be dealt with by use of the following lemma.

3.1.5 Lemma. *Suppose that f has a simple pole at a, with residue ρ. Let $\varphi_r(t) = a + re^{it}$ $(\alpha \leqslant t \leqslant \beta)$. Then*

$$\int_{\varphi_r} f \to i\rho(\beta - \alpha) \qquad \text{as } r \to 0.$$

Proof. Let $f(z) = \rho/(z-a) + g(z)$. There exist $M > 0$ and $\delta > 0$ such that $|g(z)| \leqslant M$ for $0 < |z-a| < \delta$. If $0 < r < \delta$, then $|\int_{\varphi_r} g| \leqslant M(\beta - \alpha)r$, which tends to 0 as $r \to 0$. The result follows, since $\int_{\varphi_r} [\rho/(z-a)]\,dz = i\rho(\beta - \alpha)$.

Note that the pole must be simple in 3.1.5.

Example. To evaluate $\int_{-\infty}^{\infty} (\sin x/x)\,dx$. The integrand is continuous at 0 if it is given the value 1 there. Let $f(z) = e^{iz}/z$. Then f has a simple pole at 0, with residue 1. We proceed as in the proof of 3.1.4, except that the path is modified by inserting a semicircle with centre 0 and (small) radius r in the lower half-plane. In other words, the real interval $[-r, r]$ is replaced by φ_r, where $\varphi_r(t) = re^{it}$ $(\pi \leqslant t \leqslant 2\pi)$. By 3.1.5, $\int_{\varphi_r} f \to \pi i$ as $r \to 0$. Hence we have

$$\lim_{r \to 0} \left(\int_{-\infty}^{-r} f + \int_{r}^{\infty} f \right) = \pi i.$$

Taking the imaginary part, $\int_{-\infty}^{\infty} (\sin x/x)\,dx = \pi$.

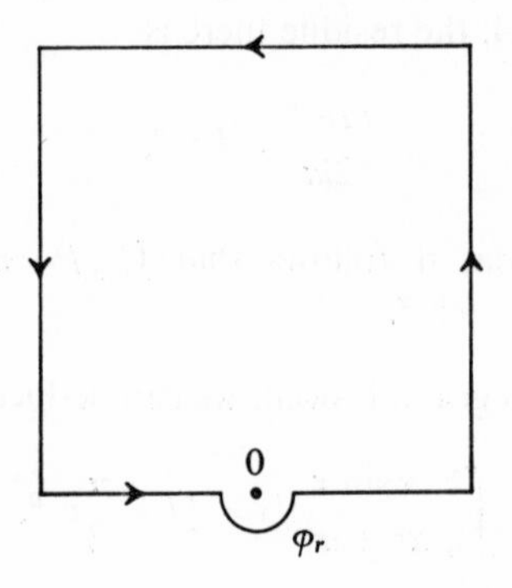

It is instructive to consider also the real part. Now $\int_r^1 (\cos x/x)\,dx \to \infty$ as $r \to 0^+$, since $\cos x/x \geqslant 1/2x$ for $0 < x < \pi/3$. However, the above shows that

$$\int_{-\infty}^{-r} \frac{\cos x}{x}\,dx + \int_{r}^{\infty} \frac{\cos x}{x}\,dx = 0$$

for each $r > 0$ (a statement which, in this case, could have been deduced immediately from the fact that the integrand is odd).

The situation here is similar to that in which $\lim_{x\to\infty} \int_{-x}^{x} f$ exists, while $\int_0^\infty f$ diverges.

The above method applies whenever there are a finite number of simple poles on the real line. The effect is to include half the residues at these poles in the sum of residues in the upper half-plane.

Non-zero contribution of the return path

With a little ingenuity, we can sometimes think of a return path on which the integral, though not small, is a simple multiple of the required integral. Such paths are usually sectors or rectangles. We illustrate this method by two examples.

Example. To evaluate $I = \int_0^\infty [x^{n-1}/(1+x^{2n})]\,dx$, where $n \geqslant 1$. Let $f(z) = z^{n-1}/(1+z^{2n})$. Then f has simple poles at $b, b^3, \ldots, b^{2n-1}$, where $b = \exp(\pi i/2n)$. Write $c = \exp(\pi i/n)$, and take $R > 0$. Let $\varphi(t) = Re^{it}$ $(0 \leqslant t \leqslant \pi/n)$, and $\psi(t) = ct$ $(0 \leqslant t \leqslant R)$. Integrate

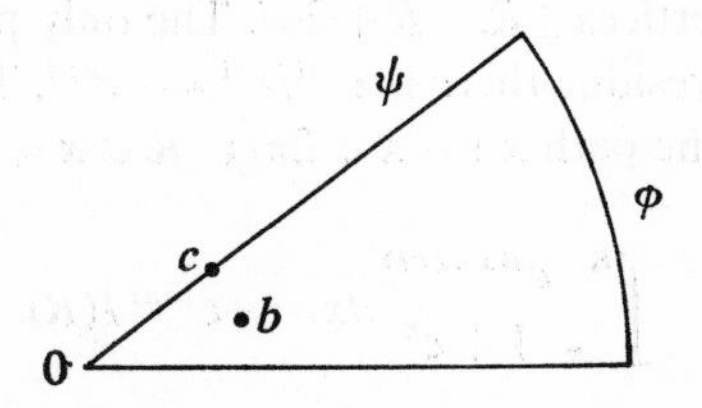

f round the sector $[0 \to R] \cup \varphi \cup (-\psi)$. This sector encloses only the pole b, where the residue is

$$\frac{b^{n-1}}{2nb^{2n-1}} = \frac{1}{2nb^n} = \frac{1}{2ni},$$

so the integral round the sector is π/n. It is elementary that $\int_\varphi f \to 0$ as $R \to \infty$, and

$$\int_\psi f = \int_0^R \frac{c^{n-1}\,t^{n-1}}{1+t^n}\,c\,dt = -\int_0^R \frac{t^{n-1}}{1+t^n}\,dt,$$

since $c^n = -1$ and $c^{2n} = 1$. It follows that $I = \pi/2n$.

It is interesting to note what would happen if we attempted to evaluate this integral by using a semicircle, as in previous examples. For n even, no result would be obtained, since the function is then odd. For n odd, the answer would be obtained, but only after the computation and addition of n residues.

Example. To evaluate $I = \int_{-\infty}^{\infty} [e^{ax}/(1+e^x)]\,dx$, where $0 < a < 1$. Let $f(z) = e^{az}/(1+e^z)$. Convergence of the integral is clear, since $0 < f(x) < e^{(a-1)x}$ $(x > 0)$ and $0 < f(x) < e^{ax}$ $(x < 0)$. Let φ be the

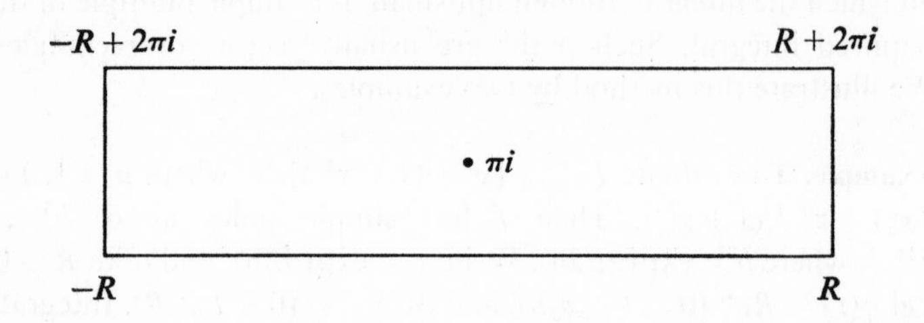

rectangle with vertices $\pm R$, $\pm R + 2\pi i$. The only pole of f inside φ is at πi, and the residue there is $e^{a\pi i}/e^{\pi i} = -e^{a\pi i}$. The top side of φ is equivalent to the path $x \mapsto x + 2\pi i$ $(-R \leqslant x \leqslant R)$. Its contribution is

$$-\int_{-R}^{R} \frac{e^{a(x+2\pi i)}}{1+e^x}\,dx = -e^{2a\pi i}\,I(R),$$

where

$$I(R) = \int_{-R}^{R} \frac{e^{ax}}{1+e^x}\,dx.$$

If z is on the right-hand side of φ, then $z = R + iy$, where $0 \leqslant y \leqslant 2\pi$, and $f(z) = e^{a(R+iy)}/(1+e^{R+iy})$. Hence $|f(z)| \leqslant e^{aR}/(e^R - 1)$, which tends to zero as $R \to \infty$. If z is on the left-hand side, then $z = -R + iy$, and $f(z) = e^{a(-R+iy)}/(1+e^{-R+iy})$. Hence $|f(z)| \leqslant e^{-aR}/(1 - e^{-R})$, which again tends to zero as $R \to \infty$. It follows that $(1 - e^{2a\pi i})I = -2\pi i e^{a\pi i}$, so

$$I = \frac{2\pi i}{e^{a\pi i} - e^{-a\pi i}} = \frac{\pi}{\sin a\pi}.$$

Note that the substitution $x = \log t$ gives

$$\int_{-\infty}^{\infty} \frac{e^{ax}}{1+e^x}\,dx = \int_0^{\infty} \frac{t^{a-1}}{1+t}\,dt.$$

Cases in which the required integral is not given by the real line

Through choice of a suitable function and path, the required integral may be given by a part of the path other than a real interval. The problem is one of recognizing that the required integral is equal to an expression of the form $\int_c^d (f \circ \varphi)\varphi'$ (or possibly to the real or imaginary part of such an expression). For instance, any integral of the form $\int_0^{2\pi} f(\cos t, \sin t)\,dt$ can be treated in this way, for, by 1.7.4, it is equal to

$$\int_{C(0,\,1)} f\left(\frac{1}{2}\left(z+\frac{1}{z}\right), \frac{1}{2i}\left(z-\frac{1}{z}\right)\right)\frac{1}{iz}\,dz;$$

(here, of course, there are no 'unwanted' bits of the path).

Example. To evaluate $I = \int_0^{2\pi} [1/(a+\cos t)]\,dt$, where $a > 1$.

$$I = \int_{C(0,\,1)} \frac{1}{iz\left(a+\frac{1}{2}\left(z+\frac{1}{z}\right)\right)}\,dz$$

$$= -2i \int_{C(0,\,1)} \frac{1}{z^2+2az+1}\,dz = -2i \int_{C(0,\,1)} \frac{1}{(z-p)(z-q)}\,dz,$$

where $p = -a + \sqrt{(a^2-1)}$, $q = -a - \sqrt{(a^2-1)}$. Now $|p| < 1$, $|q| > 1$, and the residue of the integrand at p is $1/(p-q)$, that is, $1/2\sqrt{(a^2-1)}$. Hence $I = 2\pi/\sqrt{(a^2-1)}$.

Example. To evaluate $I = \int_0^{\pi} [\cos 2t/(1-2k\cos t+k^2)]\,dt$, where $-1 < k < 1$. By substituting $t = 2\pi - u$, we see that

$$I = \frac{1}{2}\int_0^{2\pi} \frac{\cos 2t}{1-2k\cos t+k^2}\,dt.$$

Now $1-2k\cos t+k^2 = (1-ke^{it})(1-ke^{-it})$, so

$$\int_0^{2\pi} \frac{e^{2it}}{1-2k\cos t+k^2}\,dt = \int_{C(0,\,1)} f,$$

where

$$f(z) = \frac{z}{i(1-kz)\left(1-\dfrac{k}{z}\right)} = \frac{z^2}{i(1-kz)(z-k)}.$$

Now $\operatorname{res}(f,k) = k^2/i(1-k^2)$, so, taking the real part, we find that $I = \pi k^2/(1-k^2)$.

Example. Sometimes the above technique can be used to deduce an integral from another, known one. Let $I_n = \int_0^\infty x^n e^{-x}\,dx$. Integration by parts (with due attention to limits) gives $I_n = nI_{n-1}$ for $n \geqslant 1$. Since $I_0 = 1$, it follows that $I_n = n!$ for each n.

Let $f(z) = z^{4n+3}e^{-z}$, and consider the integral of f round the sector formed by Re^{it} $(0 \leqslant t \leqslant \pi/4)$ and the two radii. Cauchy's theorem shows that the total integral is 0. On the circular arc, $|f(z)| \leqslant R^{4n+3}e^{-R/\sqrt{2}}$, so this contribution tends to 0 as $R \to \infty$. The contribution of the sloping radius is

$$-\int_0^{R/\sqrt{2}} (1+i)^{4n+3}x^{4n+3}e^{-(1+i)x}(1+i)\,dx$$
$$= (-1)^n 2^{2n+2}\int_0^{R/\sqrt{2}} x^{4n+3}e^{-x}(\cos x - i\sin x)\,dx,$$

since $(1+i)^4 = -4$. Hence we have the integrals

$$\int_0^\infty x^{4n+3}e^{-x}\sin x\,dx = 0,$$

$$\int_0^\infty x^{4n+3}e^{-x}\cos x\,dx = (-1)^{n+1}\frac{(4n+3)!}{2^{2n+2}} \qquad (n = 0, 1, 2, \ldots).$$

The probability integral

Ingenuity is taken one step further in the proof of the next result, which gives the value of the important 'probability integral'. The required integral is given, not by one part of the path, but by the combination of two parallel parts.

3.1.6. $\int_{-\infty}^\infty e^{-x^2}\,dx = \sqrt{\pi}$.

Proof. We prove, in fact, that $\int_{-\infty}^\infty e^{-\pi x^2}\,dx = 1$, from which the result follows by substituting $y = x\sqrt{\pi}$. Let $f(z) = e^{i\pi z^2}$, $g(z) = f(z)/\sin \pi z$. Take $R > 0$, and write $c = e^{i\pi/4}$, so that $c^2 = i$.

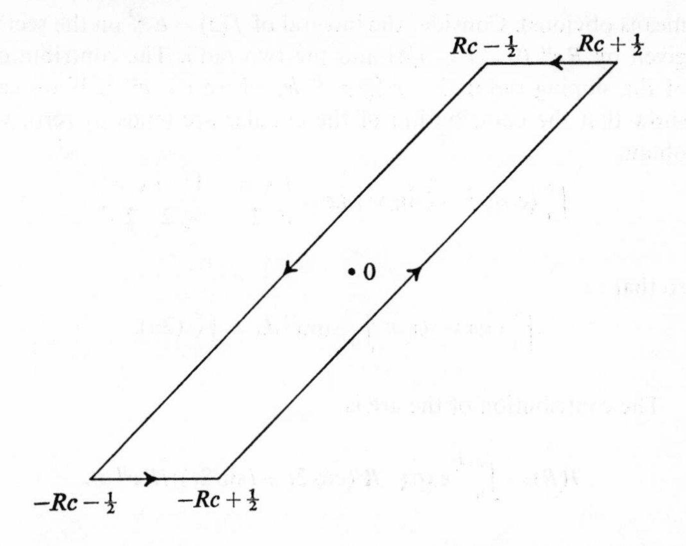

Consider the integral of g round the parallellogram φ with vertices $\pm Rc \pm \frac{1}{2}$. Now $\sin \pi(rc+\frac{1}{2}) = \cos \pi rc$ and $\sin \pi(rc-\frac{1}{2}) = -\cos \pi rc$, so the combined contribution of the sloping sides to $\int_\varphi g$ is

$$\begin{aligned}\int_{-R}^{R} \frac{1}{\cos \pi rc}(f(rc+\tfrac{1}{2}) + f(rc-\tfrac{1}{2}))\, c\, dr \\ = c\int_{-R}^{R} \exp(i\pi(ir^2+\tfrac{1}{4}))\frac{e^{i\pi rc}+e^{-i\pi rc}}{\cos \pi rc}\, dr \\ = 2c\int_{-R}^{R} c\, e^{-\pi r^2}\, dr \\ = 2i\int_{-R}^{R} e^{-\pi r^2}\, dr.\end{aligned}$$

If $z = x+iy$, then $iz^2 = -2xy + i(x^2-y^2)$, so $|f(z)| = e^{-2\pi xy}$. Also, $|\sin \pi z| \geqslant \sinh \pi|y| > 1$ for $|y| > 1$, It follows that the contributions of the horizontal sides tend to zero as $R \to \infty$.

The only zero of $\sin \pi z$ inside φ is at 0. The residue of g there is $e^0/\pi \cos 0 = 1/\pi$. Hence

$$2i\int_{-\infty}^{\infty} e^{-\pi r^2}\, dr = 2i.$$

Finally, we show how to deduce the values of $\int_0^\infty \cos x^2\, dx$ and $\int_0^\infty \sin x^2\, dx$ (notice that convergence of these integrals is by no means obvious). Consider the integral of $f(z) = e^{-z^2}$ on the sector given by Re^{it} $(0 \leqslant t \leqslant \pi/4)$ and the two radii. The contribution of the sloping radius is $-c \int_0^R e^{-ir^2}\,dr$, where $c = e^{i\pi/4}$. If we can show that the contribution of the circular arc tends to zero, we obtain

$$\int_0^\infty (\cos x^2 - i \sin x^2)\,dx = \frac{1}{c}\frac{\sqrt{\pi}}{2} = \frac{1-i}{\sqrt{2}}\frac{\sqrt{\pi}}{2},$$

so that

$$\int_0^\infty \cos x^2\,dx = \int_0^\infty \sin x^2\,dx = \tfrac{1}{4}\sqrt{(2\pi)}.$$

The contribution of the arc is

$$I(R) = \int_0^{\pi/4} \exp(-R^2(\cos 2t + i \sin 2t))\, iRe^{it}\,dt.$$

By 1.4.5,

$$|I(R)| \leqslant R \int_0^{\pi/4} e^{-R^2 \cos 2t}\,dt.$$

We use the fact that cos is concave, i.e. that $\cos t \geqslant 1 - 2t/\pi$ for $0 \leqslant t \leqslant \pi/2$. To prove this, let $g(t) = \cos t + 2t/\pi$. Then $g(0) = g(\pi/2) = 1$. Also, $g'(t) = 2/\pi - \sin t$, which decreases

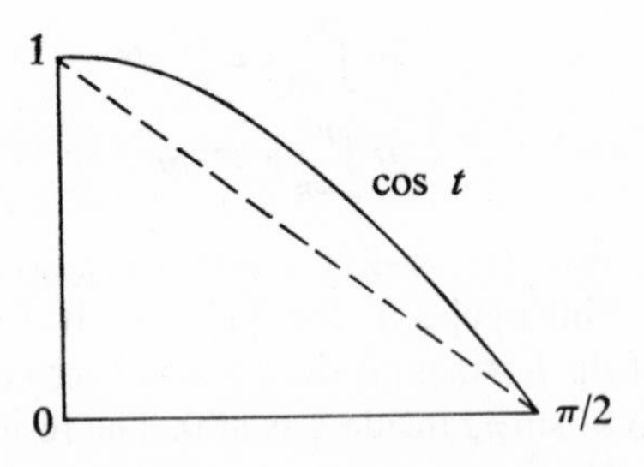

steadily from $2/\pi$ to $2/\pi - 1$ as t increases from 0 to $\pi/2$. Hence there is a unique t_0 in $(0, \pi/2)$ such that $g'(t_0) = 0$, $g'(t) > 0$ for $0 < t < t_0$, and $g'(t) < 0$ for $t_0 < t < \pi/2$. By the mean value theorem, it follows that $g(t) \geqslant g(0) = 1$ for $0 \leqslant t \leqslant t_0$ and $g(t) \geqslant g(\pi/2) = 1$ for $t_0 \leqslant t \leqslant \pi/2$. Therefore

$$\begin{aligned}|I(R)| &\leqslant R\int_0^{\pi/4} \exp\left(-R^2\left(1-\frac{4t}{\pi}\right)\right) dt \\ &= Re^{-R^2}\frac{\pi}{4R^2}(e^{R^2}-1) \\ &\leqslant \frac{\pi}{4R},\end{aligned}$$

so $I(R) \to 0$ as $R \to \infty$.

Exercises 3.1

In questions 1 to 9, show that the integrals have the values stated.

1 $\displaystyle\int_{-\infty}^{\infty} \frac{1}{1+x^4}\,dx = \pi/\sqrt{2}.$

2 $\displaystyle\int_0^{\infty} \frac{x}{1+x^4}\,dx = \frac{\pi}{4}.$

3 $\displaystyle\int_{-\infty}^{\infty} \frac{\cos \pi x}{x^2-2x+2}\,dx = -\frac{\pi}{e^{\pi}}, \quad \int_{-\infty}^{\infty} \frac{\sin \pi x}{x^2-2x+2}\,dx = 0.$

4 $\displaystyle\int_{-\infty}^{\infty} \frac{x^2}{(x^2+1)(x^2+4)}\,dx = \frac{\pi}{3}.$

5 $\displaystyle\int_{-\infty}^{\infty} \frac{1}{(1+x^2)^2}\,dx = \frac{\pi}{2}.$

6 $\displaystyle\int_{-\infty}^{\infty} \frac{x^3 \sin x}{(1+x^2)^2}\,dx = \frac{\pi}{2e}.$

7 $\displaystyle\int_{-\infty}^{\infty} \frac{x^2-a^2}{x^2+a^2}\frac{\sin x}{x}\,dx = \pi(2e^{-a}-1) \quad (a > 0).$

8 $\displaystyle\int_{-\infty}^{\infty} \frac{\cos ax - \cos bx}{x^2}\,dx = \pi(b-a) \quad (a, b \geqslant 0).$ Deduce that

$$\int_{-\infty}^{\infty} \left(\frac{\sin x}{x}\right)^2 dx = \pi.$$

9 $$\int_0^{2\pi} \frac{\cos x}{a+\cos x}\,dx = 2\pi\left(1 - \frac{a}{\sqrt{(a^2-1)}}\right) \quad (a > 1).$$

10 By considering an integral round a suitable rectangle, show that

$$\int_{-\infty}^{\infty} e^{-(x+ia)^2}\,dx = \int_{-\infty}^{\infty} e^{-x^2}\,dx$$

for all real a. Assuming that the latter is $\sqrt{\pi}$, deduce that

$$\int_{-\infty}^{\infty} e^{-x^2}\cos 2ax\,dx = e^{-a^2}\sqrt{\pi}.$$

11 By considering the integral of $z/(a-e^{-iz})$ round the rectangle with vertices $\pm\pi$, $\pm\pi + iR$, and letting $R \to \infty$, prove that

$$\int_0^{\pi} \frac{x\sin x}{1-2a\cos x + a^2}\,dx = \frac{\pi}{a}\log(1+a) \qquad (0 < a < 1).$$

12 By integrating a suitable complex function round a semicircle with the diameter indented at 0, prove that

$$\lim_{\delta\to 0+}\int_{\delta}^{\infty} \frac{\log x}{1+x^2}\,dx = 0,$$

$$\lim_{\delta\to 0+}\int_{\delta}^{\infty} \frac{\log x}{(1+x^2)^2}\,dx = -\frac{\pi}{4}.$$

(It may be assumed that $(\log x)/x \to 0$ as $x \to \infty$, but all other steps should be justified.)

3.2. The summation of series

By use of the following result, the residue theorem can be applied to find the sums of certain series of the form $\sum_{n=-\infty}^{\infty} f(n)$.

3.2.1. *Let f be a function that is differentiable on* **C** *except at a finite number of points $a_1, \ldots, a_k$ (no a_j being a real integer), and suppose that positive numbers M, R exist such that $|z^2 f(z)| \leqslant M$ whenever $|z| > R$. Let*

$$g(z) = \pi\frac{\cos \pi z}{\sin \pi z} f(z), \qquad h(z) = \frac{\pi}{\sin \pi z} f(z).$$

Then

$$\sum_{n=-\infty}^{\infty} f(n) = -\sum_{j=1}^{k} \text{res}(g, a_j),$$

$$\sum_{n=-\infty}^{\infty} (-1)^n f(n) = -\sum_{j=1}^{k} \text{res}(h, a_j).$$

Proof. Since $|f(n)| \leqslant M/n^2$ for $|n| > R$, the series $\sum f(n)$ and $\sum (-1)^n f(n)$ are convergent. The functions g and h are differentiable except at real integers and $a_1, \ldots, a_k$. If $f(n) \neq 0$, they have simple poles at the integer n, and

$$\text{res}(g, n) = \frac{\pi \cos n\pi}{\pi \cos n\pi} f(n) = f(n),$$

$$\text{res}(h, n) = \frac{\pi}{\pi \cos n\pi} f(n) = (-1)^n f(n).$$

(If $f(n) = 0$, then g and h have removable singularities at n.)

Let φ_n be the square with vertices $(n + \frac{1}{2})(\pm 1 \pm i)$. If $n > |a_j|$ for all j, then, by the residue theorem,

$$\frac{1}{2\pi i} \int_{\varphi_n} g = \sum_{j=1}^{k} \text{res}(g, a_j) + \sum_{r=-n}^{n} f(r),$$

$$\frac{1}{2\pi i} \int_{\varphi_n} h = \sum_{j=1}^{k} \text{res}(h, a_j) + \sum_{r=-n}^{n} (-1)^r f(r).$$

The result will follow if we can prove that $\int_{\varphi_n} g \to 0$ and $\int_{\varphi_n} h \to 0$ as $n \to \infty$. Take n such that $n > R$ and $n > |a_j|$ for all j. Then $|f(z)| \leqslant M/n^2$ for z in φ_n^*. Recall that for $y > 0$,

$$|\cos(x \pm iy)| \leqslant \cosh y, \qquad |\sin(x \pm iy)| \geqslant \sinh y,$$

and that $\cosh y$ and $\sinh y$ increase with y. Also,

$$\frac{\cosh y}{\sinh y} = 1 + \frac{e^{-y}}{\sinh y},$$

which decreases as y increases. If z is on one of the horizontal sides of φ_n, then $z = x \pm i(n + \frac{1}{2})$, where $-(n + \frac{1}{2}) \leqslant x \leqslant n + \frac{1}{2}$. For such z, we have

$$|\sin \pi z| \geqslant \sinh \pi > 1,$$

$$\left|\frac{\cos \pi z}{\sin \pi z}\right| \leqslant \frac{\cosh \pi}{\sinh \pi} < 2.$$

If z is on one of the vertical sides of φ_n, then $z = \pm(n+\frac{1}{2})+iy$, where $-(n+\frac{1}{2}) \leqslant y \leqslant n+\frac{1}{2}$. For such z, we have $|\sin \pi z| = \cosh \pi |y|$ and $|\cos \pi z| = \sinh \pi |y|$, so that $|\sin \pi z| \geqslant 1$ and $|\cos \pi z/\sin \pi z| \leqslant 1$. Hence $1/|\sin \pi z| \leqslant 1$ and $|\cos \pi z/\sin \pi z| \leqslant 2$ for z in φ_n^*. It follows that

$$\begin{aligned}\left|\int_{\varphi_n} g\right| &\leqslant L(\varphi_n)\sup\{|g(z)| : z \in \varphi_n^*\}\\ &\leqslant 4(2n+1)\,2\pi \frac{M}{n^2}\\ &= 8\pi M \frac{2n+1}{n^2},\end{aligned}$$

which tends to zero as $n \to \infty$. Similarly for h.

The reason for the appearance of $\sin \pi z$ in 3.2.1 is that this function has zeros at the integers. Notice that

$$\sum_{n=-\infty}^{\infty} f(n) = f(0) + \sum_{n=1}^{\infty} (f(n)+f(-n)),$$

and that if f is even, this is equal to $f(0)+2\sum_{n=1}^{\infty} f(n)$.

Example. To evaluate $S = \sum_{n=-\infty}^{\infty} [1/(a^2-n^2)]$, where a is not a real integer. Let $g(z) = (\pi \cot \pi z)/(a^2-z^2)$. Then

$$\operatorname{res}(g,a) = \operatorname{res}(g,-a) = -\frac{\pi}{2a}\cot \pi a,$$

from which we see that $S = (\pi/a)\cot \pi a$. Since the terms for n and $-n$ give the same contribution, we can deduce that

$$\sum_{n=1}^{\infty} \frac{1}{a^2-n^2} = \frac{\pi}{2a}\cot \pi a - \frac{1}{2a^2}.$$

Writing $a = ib$ (where ib is not an integer), we also have

$$\sum_{n=-\infty}^{\infty} \frac{1}{b^2+n^2} = -\frac{\pi}{ib}\cot \pi i b.$$

If b is real, it is natural to rewrite this identity in the form

$$\sum_{n=-\infty}^{\infty} \frac{1}{b^2+n^2} = \frac{\pi \cosh \pi b}{b \sinh \pi b}.$$

Example. To evaluate $\sum_{n=0}^{\infty} [(-1)^n/(2n+1)^3]$. Let

$$h(z) = \frac{\pi}{(2z+1)^3 \sin \pi z}.$$

Then h has a pole of order 3 at $-\frac{1}{2}$. We use 2.6.6 to calculate the residue there. Let $k(z) = (z+\frac{1}{2})^3 h(z) = \pi/(8 \sin \pi z)$. Then $\operatorname{res}(h, -\frac{1}{2}) = \frac{1}{2}k''(-\frac{1}{2})$. Now $k'(z) = -(\pi^2 \cos \pi z)/(8 \sin^2 \pi z)$ and $k'(-\frac{1}{2}) = 0$, so, using the fact that $\cos \pi z = \sin \pi(z+\frac{1}{2})$,

$$k''(-\tfrac{1}{2}) = -\lim_{z \to -1/2} \frac{\pi^2 \sin \pi(z+\frac{1}{2})}{8(z+\frac{1}{2}) \sin^2 \pi z} = -\frac{\pi^3}{8}$$

Hence

$$\sum_{n=-\infty}^{\infty} \frac{(-1)^n}{(2n+1)^3} = \frac{\pi^3}{16}$$

The terms for $+n$ and $-n-1$ give the same contribution, so

$$\sum_{n=0}^{\infty} \frac{(-1)^n}{(2n+1)^3} = \frac{\pi^3}{32}$$

If f has singularities at some real integers, then it is clear that 3.2.1 still holds if these integers are omitted from the sum $\sum f(n)$. As an example, we evaluate $\sum_{n=1}^{\infty} (1/n^2)$. Let $g(z) = (\pi \cos \pi z)/(z^2 \sin \pi z)$. Then g has a pole of order 3 at 0. We find that $\operatorname{res}(g, 0) = -\pi^2/3$. Hence

$$\sum_{n=1}^{\infty} \left(\frac{1}{n^2} + \frac{1}{(-n)^2}\right) = \frac{\pi^2}{3},$$

so

$$\sum_{n=1}^{\infty} \frac{1}{n^2} = \frac{\pi^2}{6}.$$

Exercises 3.2

In questions 1 to 3, prove that the series converge to the values stated.

1 $$\sum_{n=-\infty}^{\infty} \frac{1}{(2n+1)(3n+1)} = \pi/\sqrt{3}.$$

2 $$\sum_{n=-\infty}^{\infty} \frac{(-1)^n}{1+n^2} = \frac{\pi}{\sinh \pi}$$

3 $$\sum_{n=-\infty}^{\infty} \frac{1}{(n+a)^2} = \frac{\pi^2}{\sin^2 \pi a} \qquad (a \text{ not an integer}).$$

4 Evaluate

$$\sum_{n=1}^{\infty} \frac{(-1)^n}{n^2},$$

and check your answer by using the fact that

$$\sum_{n=1}^{\infty} \frac{1}{n^2} = \frac{\pi^2}{6}.$$

5 Evaluate

$$\sum_{n=1}^{\infty} \frac{1}{n(2n+1)} + \sum_{n=-1}^{-\infty} \frac{1}{n(2n+1)}$$

and check your answer by combining the terms for n and $-n$, and using a result in the text.

6 The method of this section would appear to give the obviously false conclusion that $\sum_{n=-\infty}^{\infty} e^{-n^2} = 0$. Explain this.

3.3. Partial fractions

This section begins with a result that could have been proved in 2.5. We then go on to a generalization that requires the residue theorem for its proof.

The following is no more than a watered-down (but more specific) version of 2.5.10:

3.3.1. *Suppose that f is differentiable on $\mathbf{C}$ except for a finite number of simple poles $a_1, a_2, \ldots, a_k$, and that there exist positive numbers M, R such that $|f(z)| \leqslant M$ for $|z| > R$. Then, for z different from the a_j,*

$$f(z) = c + \sum_{j=1}^{k} \frac{\operatorname{res}(f, a_j)}{z - a_j},$$

where c is a constant.

Proof. Let

$$g(z) = \sum_{j=1}^{k} \frac{\operatorname{res}(f, a_j)}{z - a_j}.$$

Then (after removing singularities) $f - g$ is an entire function. Also, $(f - g)(z)/z \to 0$ as $z \to \infty$. Therefore, by 2.3.1, $f - g$ is constant.

This expression for f is called its expression in **partial fractions.** Notice that $c = \lim_{z\to\infty} f(z)$, a fact which makes it easy to evaluate c in particular cases.

Example. Let

$$f(z) = \frac{(z+1)(z+2)(z+3)}{(z-1)(z-2)(z-3)}.$$

Then

$$\operatorname{res}(f, 1) = \frac{2.3.4}{(-1)(-2)} = 12, \qquad \operatorname{res}(f, 2) = \frac{3.4.5}{1(-1)} = -60,$$

$$\operatorname{res}(f, 3) = \frac{4.5.6}{2.1} = 60, \qquad \lim_{z\to\infty} f(z) = 1,$$

so

$$f(z) = 1 + \frac{12}{z-1} - \frac{60}{z-2} + \frac{60}{z-3}.$$

Notice that, in order to obtain this expression by a purely algebraic method, we would have to solve a system of four linear equations.

We now ask whether there is an analogous result for functions having an infinite number of simple poles $a_1, a_2, \ldots$. The simple-minded restatement of 3.3.1 would involve replacing the finite sum by $\sum_{j=1}^{\infty} [\operatorname{res}(f, a_j)/(z - a_j)]$, but unfortunately this series does not necessarily converge. However, a useful result can be obtained in the following way. Take a point z_0 different from z and the a_j, and let

$$g(\zeta) = \frac{f(\zeta)}{(\zeta - z_0)(\zeta - z)}.$$

Since the pole of f at a_j is simple, we have

$$\operatorname{res}(g, a_j) = \frac{\operatorname{res}(f, a_j)}{(a_j - z_0)(a_j - z)} = \frac{\operatorname{res}(f, a_j)}{z_0 - z}\left(\frac{1}{z - a_j} - \frac{1}{z_0 - a_j}\right).$$

Also, $\operatorname{res}(g, z) = f(z)/(z - z_0)$, and $\operatorname{res}(g, z_0) = f(z_0)/(z_0 - z)$.

Suppose that we can find a sequence $\{\varphi_n\}$ of simple, closed paths, enclosing z, z_0 and not passing through any a_j, such that $\int_{\varphi_n} g \to 0$ as $n \to \infty$. A compactness argument similar to the proof of 2.2.9 shows that only a finite number of a_j lie in any bounded set (and, in particular, inside φ_n). Let J_n be the set of j for which a_j is inside φ_n. Then the residue theorem shows that

$$f(z) = f(z_0) + \lim_{n\to\infty} \sum_{j \in J_n} \left(\frac{1}{z - a_j} - \frac{1}{z_0 - a_j}\right) \operatorname{res}(f, a_j).$$

We now state precisely a set of conditions under which this is valid.

3.3.2. *Suppose that f is differentiable on $\mathbf{C}$ except for simple poles at $a_1, a_2, \ldots$. Let $\{\varphi_n\}$ be a sequence of simple, closed, piecewise-smooth paths such that:*

(i) *$a_j \notin \varphi_n^*$ for all j, n;*

(ii) *there exist α, $\beta > 0$ such that, for each n, $L(\varphi_n) \leqslant \alpha n$ and $|\zeta| \geqslant \beta n$ for all ζ in φ_n^*;*

(iii) *there exists M such that $|f(\zeta)| \leqslant M$ for ζ in $\bigcup \varphi_n^*$.*

Let J_n be the set of j for which a_j is inside φ_n. If z, z_0 are distinct points that are different from all the a_j, then

$$f(z) = f(z_0) + \lim_{n\to\infty} \sum_{j \in J_n} \left(\frac{1}{z - a_j} - \frac{1}{z_0 - a_j}\right) \operatorname{res}(f, a_j).$$

Proof. Defining g as above, it is sufficient to show that $\int_{\varphi_n} g \to 0$ as $n \to \infty$. Let $k = \max(|z|, |z_0|)$. If $n > 2k/\beta$ and $\zeta \in \varphi_n^*$, then $|\zeta| > \beta n > 2k$, so

$$|g(\zeta)| \leqslant \frac{M}{\frac{1}{4}|\zeta|^2} \leqslant \frac{4M}{\beta^2 n^2}.$$

Hence

$$\left|\int_{\varphi_n} g\right| \leqslant \frac{4M}{\beta^2 n^2}\alpha n,$$

which tends to zero as $n \to \infty$.

If f has a simple pole at 0, we can apply 3.3.2 to the difference between f and its principal part at 0 to obtain the following variant of the theorem, in which z_0 is 0, despite the fact that 0 is a pole. This form of the result is often useful.

3.3.3 Corollary. *Suppose that f is differentiable on* $\mathbf{C}$ *except for simple poles at* $0, a_1, a_2, \ldots$. *Let* $\{\varphi_n\}$ *be a sequence of simple, closed, piecewise-smooth paths satisfying conditions* (i), (ii), (iii) *of* 3.3.2, *and let J_n be the set of j for which a_j is inside φ_n. Let the Laurent series for f at 0 be $\sum_{n=-1}^{\infty} b_n z^n$. Then, for z different from 0 and the a_j,*

$$f(z) = \frac{b_{-1}}{z} + b_0 + \lim_{n\to\infty} \sum_{j\in J_n} \left(\frac{1}{z-a_j} + \frac{1}{a_j}\right) \operatorname{res}(f, a_j).$$

Proof. Let $f_1(z) = f(z) - b_{-1}/z$ for $z \neq 0$, and $f_1(0) = b_0$. Then f_1 is differentiable at 0, and has a simple pole at a_j, with residue equal to $\operatorname{res}(f, a_j)$. Also, f_1 is bounded on $\bigcup \varphi_n^*$. The result follows, by applying 3.3.2 to f_1, with $z_0 = 0$.

Example. The function $z \mapsto \pi \cot \pi z$ has simple poles at the integers, the residue being 1 in each case. Since the function is odd, the constant term in its Laurent series at 0 vanishes. Let φ_n be the square with vertices $(n+\frac{1}{2})(\pm 1 \pm i)$. As shown in the proof of 3.2.1, $|\cot \pi z| \leqslant 2$ for z in φ_n^* $(n > 1)$. Hence, if z is not an integer,

$$\begin{aligned}\pi \cot \pi z &= \frac{1}{z} + \lim_{n\to\infty} \sum_{r=1}^{n} \left[\left(\frac{1}{z-r} + \frac{1}{r}\right) + \left(\frac{1}{z+r} - \frac{1}{r}\right)\right] \\ &= \frac{1}{z} + \lim_{n\to\infty} \sum_{r=1}^{n} \left(\frac{2z}{z^2 - r^2}\right),\end{aligned}$$

so

$$\pi \cot \pi z = \frac{1}{z} + 2z \sum_{n=1}^{\infty} \frac{1}{z^2 - n^2}.$$

This is exactly the identity obtained in the example following 3.2.1. In 3.2, our aim was to express the sum of the series in terms of the 'known' function cot; in the present section, we have found (within limits) how this process can be reversed.

The above also yields the following pleasing identity:

$$\pi \cot \pi z = \lim_{n\to\infty} \sum_{r=-n}^{n} \frac{1}{z-r}.$$

Corresponding expressions for $\tan \pi z$ can be obtained similarly, or by using the fact that $\tan \pi z = -\cot \pi(z - \frac{1}{2})$.

We have restricted our considerations to functions whose poles are all simple. As the reader may have suspected, this is not the end of the partial-fractions story. For further material on the subject, we refer to Ahlfors [1], pages 185–8, or Duncan [2], pages 214–16.

Exercises 3.3

1 Express the following functions in partial fractions:

$$\frac{(z+1)^2}{(z-1)(z+2)(z+4)}, \quad \frac{(z-2)(2z+1)(3z+1)}{z(z^2-1)}.$$

2 Prove that, for z not an integer,

$$\frac{\pi}{\sin \pi z} = \frac{1}{z} + 2z \sum_{n=1}^{\infty} \frac{(-1)^n}{z^2 - n^2}.$$

3 Prove that

$$\frac{1}{e^z - 1} = \frac{1}{z} - \frac{1}{2} + 2z \sum_{n=1}^{\infty} \frac{1}{z^2 + 4n^2\pi^2}.$$

4 In 3.3.2, prove that convergence is uniform on any bounded, closed set that does not contain z_0 or any a_j.

5 Write down the series obtained for $\pi \cot \pi z$ by putting $z_0 = i$ in 3.3.2.

3.4. Winding numbers

Winding numbers were given a very summary treatment in 2.6. In the present section we come back to take a closer look at them.

First we show that winding numbers are always integers. The following proof should be compared with 2.2.6.

3.4.1. *Let φ be a piecewise-smooth path from a to b, and let z_0 be a point not in φ^*. Let $v = \int_\varphi [1/(z - z_0)]\,dz$. Then*

$$e^v = \frac{b - z_0}{a - z_0}.$$

Proof. Let $[c,d]$ be the domain of φ. For t in $[c,d]$, let

$$h(t) = \int_c^t \frac{\varphi'(u)}{\varphi(u) - z_0}\,du,$$

i.e. the integral of $1/(z - z_0)$ along φ as far as $\varphi(t)$. Then, by 1.4.2, $h'(t) = \varphi'(t)/(\varphi(t) - z_0)$, so the derivative of $(\varphi(t) - z_0)e^{-h(t)}$, that is,

$$e^{-h(t)}\{\varphi'(t) - h'(t)(\varphi(t) - z_0)\},$$

is zero for $c \leqslant t \leqslant d$. It follows that

$$(\varphi(d) - z_0)\,e^{-h(d)} = (\varphi(c) - z_0)\,e^{-h(c)}.$$

This gives the result, since $h(c) = 0$, $h(d) = v$, $\varphi(c) = a$, $\varphi(d) = b$.

Using 1.6.11, we can now deduce at once:

3.4.2 Corollary. *If φ is a closed, piecewise-smooth path, and z_0 is not in φ^*, then $\mathrm{w}(\varphi, z_0)$ is an integer.*

3.4.3. *Let φ be a closed, piecewise-smooth path, and let G be a connected, open set containing no point of φ^*. Then there is an integer n such that $\mathrm{w}(\varphi, z) = n$ for all z in G.*

Proof. We show that the mapping $z \mapsto \mathrm{w}(\varphi, z)$ is continuous on G. The result then follows, for if $\mathrm{w}(\varphi, z) = n$ for some z in G, then $\{z \in G : \mathrm{w}(\varphi, z) = n\}$ is a non-empty set that is both open and closed in G.

Take z_0 in G and ϵ in $(0, \frac{1}{2})$. There exists $r > 0$ such that $D(z_0, r)$ is contained in G, and so does not meet φ^*. Suppose that $|z - z_0| \leqslant r\epsilon$. Then, for ζ in φ^*, $|\zeta - z| \geqslant r/2$, so

$$\left|\frac{1}{\zeta - z} - \frac{1}{\zeta - z_0}\right| = \frac{|z - z_0|}{|\zeta - z||\zeta - z_0|} \leqslant \frac{\epsilon r}{\frac{1}{2}r^2} = \frac{2\epsilon}{r}.$$

Hence $|\mathrm{w}(\varphi, z) - \mathrm{w}(\varphi, z_0)| \leqslant (\epsilon/\pi r)L(\varphi)$.

Geometrical interpretation

An alternative way of dealing with winding numbers is to break up the path into pieces each of which lies in a star-shaped set not containing z_0, and to apply the results on logarithms proved in 1.6. As well as giving another proof of 3.1.2, this approach gives us an intuitive idea of the meaning of winding numbers, and a method of computing them in particular cases.

If $z_0 \in \mathbf{C}$ and $\alpha \in \mathbf{R}$, let

$$H_\alpha(z_0) = \{z_0 - r\,e^{i\alpha} : r \geqslant 0\}.$$

This is a half-line starting at z_0; with the notation of 1.6, it is $z_0 + H_\alpha$. By 1.6.16, $(d/dz)\log_\alpha(z - z_0) = 1/(z - z_0)$ for z in $\mathbf{C}\setminus H_\alpha(z_0)$. Hence, by 1.7.5, we have:

3.4.4. *If ψ is a piecewise-smooth path in $C\setminus H_\alpha(z_0)$, with initial point a and final point b, then*

$$\int_\psi \frac{1}{z - z_0}\,dz = \log_\alpha(b - z_0) - \log_\alpha(a - z_0)$$

$$= \log\frac{|b - z_0|}{|a - z_0|} + i(\arg_\alpha(b - z_0) - \arg_\alpha(a - z_0)).$$

Let φ be a closed, piecewise-smooth path, and let z_0 be a point not in φ^*. Then φ can be expressed in the form $\varphi_1 \cup \cdots \cup \varphi_k$, where each φ_j lies in a set of the form $\mathbf{C}\setminus H_{\alpha_j}(z_0)$ (we omit the proof of this statement, which is a routine application of compactness). Let a_{j-1}, a_j be the initial and final points of φ_j (so that $a_k = a_0$). Then, by 3.4.4,

$$\int_\varphi \frac{1}{z - z_0}\,dz = i\sum_{j=1}^{k} [\arg_{\alpha_j}(a_j - z_0) - \arg_{\alpha_j}(a_{j-1} - z_0)]$$

$$= i\sum_{j=1}^{k} (\arg_{\alpha_{j+1}} - \arg_{\alpha_j})(a_j - z_0)$$

$$+ i(\arg_{\alpha_k} - \arg_{\alpha_1})(a_0 - z_0).$$

Since arguments of a complex number differ by an integer multiple of 2π, this shows that $\int_\varphi [1/(z-z_0)]\,dz = 2n\pi i$ for some integer n. If we start by selecting an argument of $a_0 - z_0$, and then choose arguments $\theta(t)$ of $\varphi(t) - z_0$ $(c \leqslant t \leqslant d)$ so that θ varies continuously with t, then, clearly, $2n\pi = \theta(d) - \theta(c)$.

We can now give the promised intuitive idea of the meaning of winding numbers. (In the following, we are not proving the equality of two mathematically defined quantities, but equating one mathematically defined quantity with a fairly obvious physical notion.) Imagine an elastic string with one end fixed at z_0 and the other moving with $\varphi(t)$ as t increases from c to d. Each increase of 2π in the function θ defined above evidently corresponds to a complete rotation of the string in an anti-clockwise direction. It is therefore reasonable to say that $\mathrm{w}(\varphi, z_0)$ is the number of times φ goes round z_0 in the anti-clockwise sense.

An example

Consider the path $\varphi(t) = 4e^{it}\cos\frac{2}{3}t$ $(0 \leqslant t \leqslant 6\pi)$. By breaking up the path into convenient pieces, we show that $\mathrm{w}(\varphi, 1) = 3$.

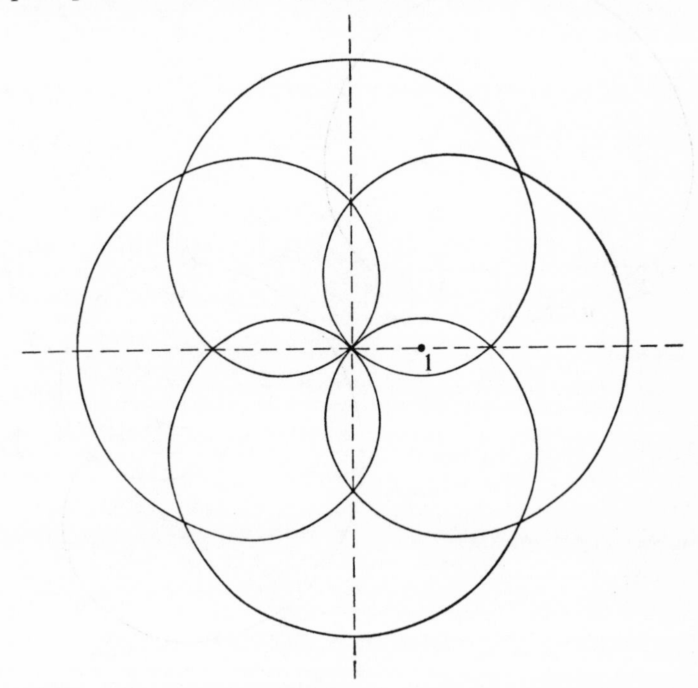

First we consider the sub-interval $[0, \frac{3}{4}\pi]$ of $[0, 6\pi]$. For $0 < t < \frac{3}{4}\pi$, $\sin t > 0$ and $\cos \frac{2}{3}t > 0$, so $\operatorname{Im} \varphi(t) > 0$. Also, $\varphi(0) = 4$ and $\varphi(\frac{3}{4}\pi) = 0$. Hence this part of the path lies in the upper half-plane, and does not meet $\{1 - ri : r \geqslant 0\} = H_{\pi/2}(1)$. Therefore its contribution to $\operatorname{Im} \int_\varphi [1/(z-1)]\,dz$ is

$$\arg_{\pi/2}(-1) - \arg_{\pi/2}(3) = \pi - 0 = \pi.$$

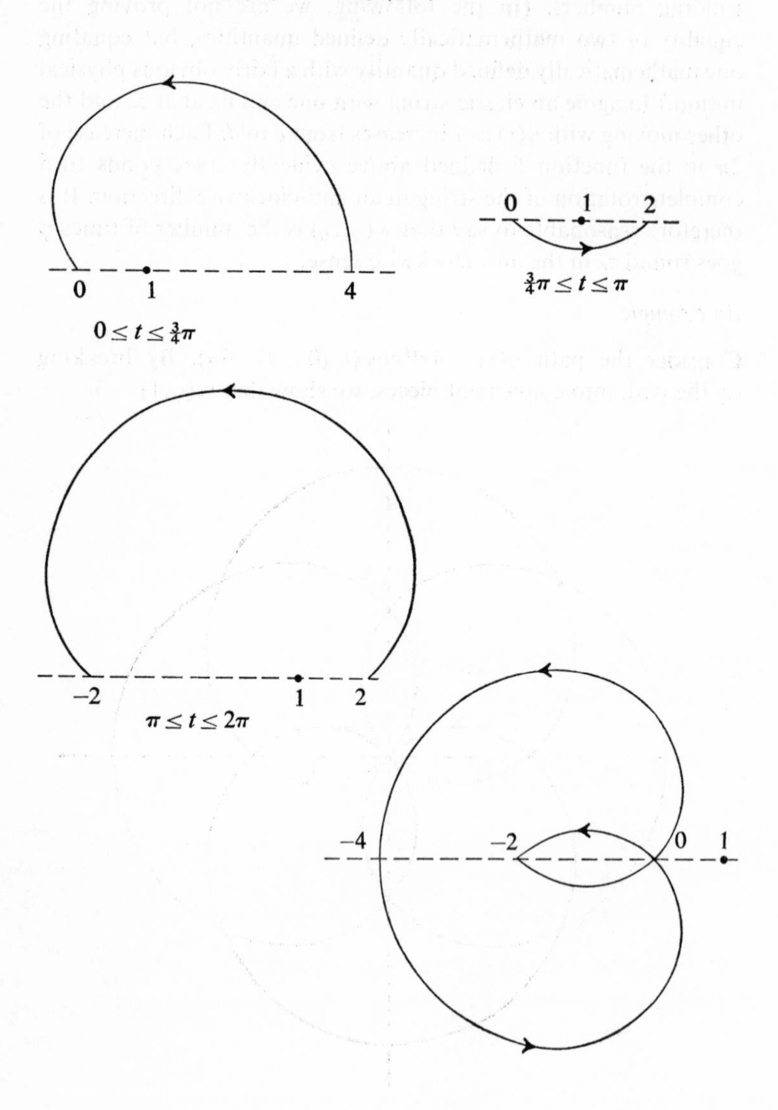

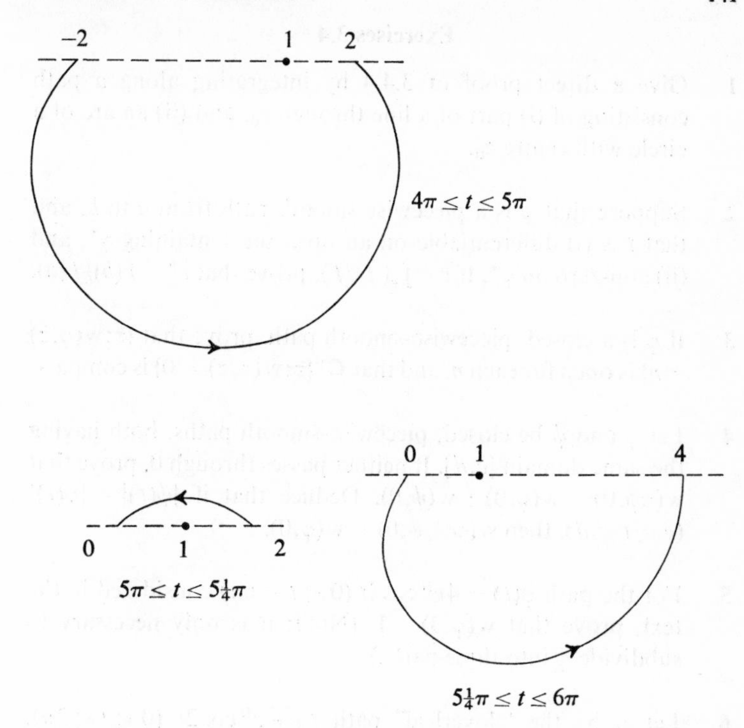

Similarly, we can show that each of the following sub-intervals of $[0, 6\pi]$ contributes π to $\operatorname{Im} \int_\varphi [1/(z-1)]\,dz$:

$$[\tfrac{3}{4}\pi, \pi], \qquad [\pi, 2\pi], \qquad [4\pi, 5\pi], \qquad [5\pi, 5\tfrac{1}{4}\pi], \qquad [5\tfrac{1}{4}\pi, 6\pi].$$

Finally, we show that the contribution of $[2\pi, 4\pi]$ is zero. Now $\varphi(2\pi) = \varphi(4\pi) = -2$, so this part of the path is closed. We show that it does not meet $\{r : r \geqslant 1\}$, from which the assertion follows. The only values of t in $[2\pi, 4\pi]$ for which $\varphi(t)$ is non-zero and $\operatorname{Im} \varphi(t) = 0$ are $2\pi, 3\pi, 4\pi$. Since $\varphi(3\pi) = -4$, the required property holds.

It is important to realize that the diagrams do not constitute the proof in this example, but only assist understanding of the proof.

Exercises 3.4

1 Give a direct proof of 3.4.4 by integrating along a path consisting of (i) part of a line through z_0, and (ii) an arc of a circle with centre z_0.

2 Suppose that φ is a piecewise-smooth path from a to b, and that f is (i) differentiable on an open set containing φ^*, and (ii) non-zero on φ^*. If $v = \int_\varphi (f'/f)$, prove that $e^v = f(b)/f(a)$.

3 If φ is a closed, piecewise-smooth path, prove that $\{z : \mathrm{w}(\varphi, z) = n\}$ is open for each n, and that $\mathbf{C} \setminus \{z : \mathrm{w}(\varphi, z) = 0\}$ is compact.

4 Let φ and ψ be closed, piecewise-smooth paths, both having the same domain $[c, d]$. If neither passes through 0, prove that $\mathrm{w}(\varphi\psi, 0) = \mathrm{w}(\varphi, 0) + \mathrm{w}(\psi, 0)$. Deduce that if $|\psi(t)| < |\varphi(t)|$ $(c \leqslant t \leqslant d)$, then $\mathrm{w}(\varphi + \psi, 0) = \mathrm{w}(\varphi, 0)$.

5 For the path $\varphi(t) = 4e^{it}\cos\frac{2}{3}t$ $(0 \leqslant t \leqslant 6\pi)$ considered in the text, prove that $\mathrm{w}(\varphi, 3) = 1$. (Note: it is only necessary to subdivide φ into three parts.)

6 Let φ be the 'cloverleaf' path $t \mapsto e^{it}\cos 2t$ $(0 \leqslant t \leqslant 2\pi)$. By considering the point re^{it_0}, where (i) $0 < r < |\cos 2t_0|$, (ii) $r > |\cos 2t_0|$, prove that φ is simple.

Glossary of symbols

Symbol	Meaning	
Re	real part	*page* 10
Im	imaginary part	10
$\vert\ \vert$	modulus	11
$\bar{z}$	conjugate	11
$D(a, r)$	disc	12
$D'(a, r)$	punctured disc	12
$[a:b]$	line segment	20
co( )	convex cover	21
φ^*	range of path	26
$-\varphi$	reverse path	26
$\varphi \cup \psi$	join of paths	27
$[a \to b]$	directed line segment	27
$L(\ )$	length of path	28
exp	exponential	39
cos	cosine	39
sin	sine	39
$C(a, r)$	circle	44
$\arg_\alpha$		46

$\log_\alpha$		47
$\int_\varphi f$	integral on φ	54
∂T	perimeter of triangle T	63
$\mathrm{ord}\,(f, z_0)$	order of f at z_0	95
$\mathrm{res}\,(f, z_0)$	residue	101
$\mathrm{w}\,(\varphi, z)$	winding number	101
$\mathrm{ZP}\,(f, \varphi)$	number of zeros minus number of poles	107

Bibliography

Further reading on Complex Analysis

[1] AHLFORS, L. V. (1966). *Complex Analysis*, 2nd ed., McGraw-Hill, New York.

[2] DUNCAN, J. (1968). *The Elements of Complex Analysis*, Wiley, New York.

[3] MACKEY, G. W. (1967). *Lectures on the Theory of Functions of a Complex Variable*, Van Nostrand, New York.

[4] NEVANLINNA, R. and PAATERO, V. (1969). *Introduction to Complex Analysis*, Addison-Wesley, New York.

Real Analysis

[5] MOSS, R. M. F. and ROBERTS, G. T. (1968). *A Preliminary Course in Analysis*, Chapman and Hall, London.

Metric spaces

[6] SIMMONS, G. F. (1963). *An Introduction to Topology and Modern Analysis*, McGraw-Hill, New York.

Functions on $\mathbf{R}^n$

[7] SPIVAK, M. (1965). *Calculus on Manifolds*, Benjamin, New York.

Index